THE
PAUL HAMLYN
LIBRARY
DONATED BY
THE PAUL HAMLYN
FOUNDATION
TO THE
BRITISH MUSEUM
opened December 2000
WITHDRAWN

ANTIQUARIAN HOROLOGICAL SOCIETY MONOGRAPH NO. 25

THE HALL OF HEAVENLY RECORDS

李約瑟 康必知
魯桂珍 馬絳 著

朝鮮「書雲觀」天文儀器與計時機

周士一書

周士一印

THE HALL OF HEAVENLY RECORDS

KOREAN ASTRONOMICAL INSTRUMENTS AND CLOCKS 1380–1780

JOSEPH NEEDHAM
LU GWEI-DJEN
JOHN H. COMBRIDGE
JOHN S. MAJOR

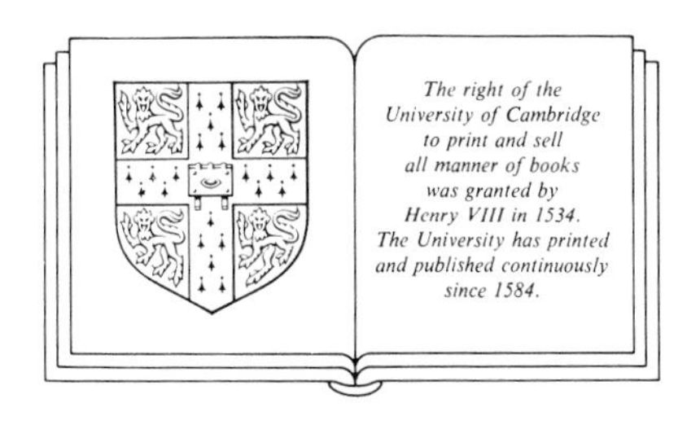

CAMBRIDGE UNIVERSITY PRESS

CAMBRIDGE

LONDON NEW YORK NEW ROCHELLE

MELBOURNE SYDNEY

Published by the Press Syndicate of the University of Cambridge
The Pitt Building, Trumpington Street, Cambridge CB2 1RP
32 East 57th Street, New York, NY 10022, USA
10 Stamford Road, Oakleigh, Melbourne 3166, Australia

Monograph No. 25
Antiquarian Horological Society,
New House, High Street, Ticehurst,
Wadhurst, Sussex TN5 7AL

First published 1986

Printed in Great Britain by the University Press, Cambridge

Library of Congress catalogue card number: 85-4229

British Library cataloguing in publication data
The Hall of Heavenly Records:
Korean astronomical instruments and clocks 1380–1780.
1. Scientific apparatus and instruments – Korea – History
I. Needham, Joseph
522′.2′09519 Q185.7

ISBN 0 521 30368 0

CONTENTS

LIST OF ILLUSTRATIONS

A NOTE ON ROMANISATION AND OTHER CONVENTIONS

Names of Chinese persons, places, institutions, etc. are given in Wade–Giles romanisation, with the exception of a few familiar place-names that are widely recognised in older forms (Peking, Kaifeng, etc.). Korean romanisation follows the McCune–Reischauer system. In accordance with recently accepted standard practice, Korean personal names are not hyphenated, except in the case of Korean authors of works in English who hyphenate their own names. At the first appearance in each chapter of any Korean word or phrase we provide, in parentheses, its Chinese characters; except in the case of proper nouns, we also provide Wade–Giles Chinese romanisation for Korean terms. Words singled out for special attention in our translations are given in Chinese romanisation only, as the Korean historical annals are written in standard Classical Chinese. Characters and dates for the kings of the Yi Dynasty are given in the Appendix.

Dates are denoted as Before Common Era (B.C.E.) and Common Era (C.E.).

Where it is possible to specify exact dimensions in our discussions of instruments and the like, we employ metric units. The words 'feet' and 'inches', both in the translations and in our own text, refer to the Chinese ('Chou') foot (*ch'ih* 尺) and inch (*ts'un* 寸, i.e. $\frac{1}{10}$ of a *ch'ih*).

Within translated passages, words clearly implied by but not present in the original text, or that have been supplied to meet the requirements of English grammar and syntax, are enclosed in parentheses. Words that convey the translators' comments, amplifications, or other interpolations are enclosed in square brackets.

INTRODUCTION AND ACKNOWLEDGEMENTS

In this book we present a study of the astronomical instruments and star-charts of Korea – the Kingdom of Chosŏn 朝鮮 – during the greater part (*c.* 1392–1776) of the Yi 李 Dynasty. The focus of our study is the Korean Royal Observatory and Bureau of Astronomy, the Sŏun Kwan (Shu-yün Kuan 書雲觀, 'The Watch-tower for Recording Celestial Ephemera'), which we have styled 'The Hall of Heavenly Records'.[1] We have confined our investigations to the astronomical and horological instruments, celestial planispheres, and other physical paraphernalia of the Sŏun Kwan; we have not attempted a history of Korean observational astronomy and meteorology during the period in question, nor do we deal extensively with political and institutional concerns.

The materials for this study are drawn from two types of evidence: first a small but important array of physical objects, in the form of a few instruments and star-maps that have survived the ravages of warfare and time down to the present day, and secondly documentary evidence. The latter is found mainly in the 'Veritable Records' (*sillok*; *shih-lu* 實錄) of the various kings of the Yi Dynasty[2] and in a great Korean historical work, the *Chŭngbo munhŏn pigo* 增補文獻備考, the 'Comprehensive Study of Civilisation, Revised and Expanded'.[3] In what follows, we

[1] The name of this bureau was changed in 1466 to Kwansanggam (Kuan-hsing-chien 觀星監, 'Superintendency of the Observation of Celestial Phenomena'). See Jeon Sang-woon, *Science and Technology in Korea: Traditional Instruments and Techniques* (Cambridge, Mass. and London: MIT Press, 1974), p. 105. (Hereafter Jeon, STK.)

[2] The 'Veritable Records' of the Yi kings, collectively usually called the *Yijo sillok* 李朝實錄 or the *Chosŏn wangjo sillok* 朝鮮王朝實錄, were compiled from contemporary primary sources by special temporary bureaus set up after the death of each king. Some kings do not have 'Veritable Records', but rather 'diaries' (*ilgi*; *jih-chi* 日記). For various editions of the Veritable Records, see Tu-jong Kim, *A Bibliographical Guide to Traditional Korean Sources* (Seoul: Asiatic Research Centre, Korea University, 1976), pp. 61 ff, and Benjamin H. Hazard *et al.*, *Korean Studies Guide* (Berkeley: University of California Press, 1954), pp. 100–1. We have used a modern Japanese reprint edition, *Richō jitsuroku* 李朝實錄 (Tokyo: Gakushūin Institute of Oriental Culture, 1953).

[3] Officially compiled; published 1790, 1908. We have used a modern reprint edition published in Seoul by the Kosŏ Kan-haenghoe 古書刊行會, 1959.

present translations of long passages from these works relating to astronomical instruments and the like, and also (in Chapter 5) translations of inscriptions on an important surviving Korean star-map of the middle eighteenth century. Through a presentation and analysis of this evidence we have tried to identify and describe as precisely as possible the instruments of the Korean Royal Observatory, to explain their functions, and to show how they fit into the larger context of East Asian astronomy.

The question of context is an important one, for Korean astronomy was continually and powerfully influenced by that of its great neighbour China. Most of the instruments and other materials that we discuss below are, directly or indirectly, of Chinese ancestry. Accordingly, we have tried wherever possible to trace that line of descent, paying particular attention to the instruments of the great Astronomer Royal of the Chinese Yüan Dynasty, Kuo Shou-ching 郭守敬 (*c.* 1280 C.E.), and to the works of the Jesuit astronomers of the Chinese Mission at Peking, whose Shih-hsien 時憲 calendar (adopted by the Ch'ing Dynasty in 1645 and by the Yi court in 1651), along with their other astronomical writings, wrought important changes in the theoretical basis of Sino-Korean astronomy. The question of context also bears on the pace and timing of the activities of the Korean Royal Observatory. Because of Korea's political status as a tributary of the Chinese empire, the two greatest eras of astronomical instrument-building in the Yi Period (the early fifteenth and middle seventeenth centuries) were to an important extent responses to the consequences of dynastic overthrow and renewal in China, as well as being evoked by circumstances that were more directly confined to Korea itself.

The plan of this work is as follows:

In Chapter 1 we present a brief introduction to the theoretical background to Chinese and Sino-Korean astronomy, discussing cosmological, calendrical, horological, and mechanical matters that are essential to an understanding of what follows. We then briefly discuss the historical background to the first great period of astronomical instrument-making in the early Yi Dynasty, namely the fall of the Mongol Yüan Dynasty of China, the subsequent fall of the Korean Kingdom of Koryŏ 高麗, and the reigns of the three great early Yi kings, T'aejo 太祖, T'aejong 太宗, and Sejong 世宗.[4]

In Chapter 2 we describe and discuss the instruments made by order of King

[4] The (posthumous) names and reign-dates of the Yi Dynasty kings are given in the Appendix.

Sejong in his re-equipping of the Royal Observatory in the 1430s. These included, among others, armillary spheres,[5] copies of Kuo Shou-ching's Simplified Instrument (the 'equatorial torquetum'), a variety of sundials, and two complex mechanical clepsydras.[6] We present and comment upon translations of descriptions of the instruments from the *Sejong sillok* and the *Chŭngbo munhŏn pigo*, and discuss the operating principles of the instruments; where possible, we present as well photographs of surviving instruments, or reconstructional drawings based on the literary descriptions.

In Chapter 3 we discuss the fate of King Sejong's instruments, as they were repaired, replaced, or augmented in succeeding reigns. Most of them were destroyed in 1592, when Korea was invaded by Japanese armies under the great general Hideyoshi. The task of replacing them was slowed by further invasions of Korea, this time by the Manchus, in the early and middle seventeenth century; but after the Korean adoption of the Ch'ing (Jesuit-inspired) Shih-hsien calendar in 1651, a new burst of instrument-making activity occurred. In the latter part of this chapter we discuss the making of several important horological instruments in the 1660s, and then follow the aftermath of this story through the reign of King Yŏngjo 英祖 (to 1776).

Chapter 4 presents a detailed technical description of the most important, and only surviving, horological instrument among those from the reign of King Hyŏnjong 顯宗 described in Chapter 3, namely the 1669 armillary clock of Song Iyŏng 宋以穎 incorporating an armillary sphere of Yi Minch'ŏl 李敏哲.

Chapter 5 (a revised version of a previously published article by Needham and Lu) presents a similarly detailed description of a Korean astronomical screen of the mid eighteenth century. This chapter provides us with a further opportunity for

[5] Armillary spheres are nests of rings which represent various significant great circles of the celestial sphere (the equator, ecliptic, meridian, horizon, etc.). In these pages we shall encounter two types: *observational* armillary spheres, designed to be used (usually with a sighting alidade or tube) to plot the locations and movements of the heavenly bodies; and *demonstrational* armillary spheres, designed simply to model the heavens for the contemplation of onlookers. In either case, the armillary sphere was usually designed so as to allow various rings to be rotated about an axis, either manually or mechanically. See Joseph Needham, *Science and Civilisation in China*, vol. III (Cambridge: Cambridge University Press, 1959), pp. 342–54 *et seq.* (Hereafter Needham, SCC.) See also below, Ch. 2, n. 11, on pp. 20–1.

[6] A clepsydra is any timekeeper operated by means of a regulated flow of liquid into ('inflow type') or out from ('outflow type') a vessel where it is measured. The clepsydras discussed in these pages are of the inflow type. Special kinds that will be encountered here include the *float* or *float-rod* clepsydra, in which a float in the inflow vessel carries a visible rod marked with divisions of time, and the *anaphoric* clepsydra, in which a string from the float imparts rotary motion to an axle or drum. Either type could be used to transmit power to the mechanisms of a *striking* clepsydra to operate audible (and sometimes also visible) time signals.

discussing the background of Chinese Jesuit astronomy and its influence on Korea.

We conclude our book with an epilogue that touches briefly on events in the last part of the Yi period.

The authors hope that this study of the instruments of the Korean Royal Observatory will lead others to take up an investigation of the many unresolved questions, including the history of observational records,[7] that still remain in the field of Korean astronomy. The bibliography is designed as an aid to those who may wish to pursue this interesting and rewarding task; for their benefit it includes some entries not directly mentioned in our text and footnotes.

In a work published in 1905, an ardent friend of the Korean people made the following observation: 'In 1550 ... an astronomical instrument was made, called the ... "Heaven Measure". We are not told the exact nature of the instrument, but it implies a considerable degree of intellectual activity and an inclination toward scientific pursuits that is rare in Korea.'[8] This assertion could hardly be more mistaken. It reflects prejudices to which no responsible scholar today could subscribe, but which have died a most reluctant and lingering death: namely that East Asia has produced little science worthy of the name, and that Korean science has been but a pale reflection of that of the Chinese. The great flowering of historical studies of Chinese science in recent years has done much to eradicate the first prejudice, while the great pioneers of the study of traditional Korean science, Carl Rufus and his collaborators and (a generation later) Jeon Sang-woon, have done much to demolish the second.[9]

Our study of the instruments of the Korean Royal Observatory, which builds on the labours of Rufus and Jeon, has convinced us that while Korean astronomy

[7] Some Korean observational records, mostly those in the official histories, have already been used by Chinese scholars; see for example the article by Hsi Tse-tsung (Xi Zezong) 席泽宗 and Po Shu-jen (Bo Shuren) 薄树人, 'Chung Ch'ao Jih san-kuo ku-tai ti hsin-hsing chi-lu chi ch'i tsai she-tien t'ien-wen-hsüeh chung ti i-i' 中朝日三国古代的新星纪录及其在射电天文学中的意义, *T'ien-wen hsüeh-pao* 天文學報 (*Acta Astronomia Sinica*), 1965, *13*.1: 1–22, tr. as S. R. Bo and Z. Z. Xi, 'Ancient Novae and Supernovae Recorded in the Annals of China, Korea and Japan and their Significance in Radio Astronomy', NASA TT-F-388 (Technical Translations Series), 1966. Other observational records probably remain in MS. from among the Korean government archives, and we may hope that someday they will become available to scholars.

[8] Homer B. Hulbert, *History of Korea*, 2 vols. (1905; repr. ed. Clarence N. Weems, 2 vols., London, 1962), 1: 334.

[9] See the publications of Rufus, and Jeon, listed in our bibliography.

was indeed based solidly on that of China, it also had important indigenous features, wrought significant changes upon Chinese ideas and techniques, and incorporated influences from the Peking Jesuits (and, to a lesser extent, from Japan) in ways that sometimes differed from the acceptance of those influences in China. Korean astronomy was, in other words, a true national variant of the East Asian astronomical tradition; the instruments and written records that it produced are a valuable legacy to the history of science everywhere.

We think it appropriate, therefore, to consider here some of the implications of Hulbert's statement that 'an inclination toward scientific pursuits ... is rare in Korea'. He may have been thinking of the passing situation in his own time, but it is fair to say that modern historians of East Asian science have come to a precisely opposite conclusion. Twice in *Science and Civilisation in China* the writers were moved to say that 'of all the peoples on the periphery of the Chinese culture-area, the Koreans were probably the most interested in science, mechanical technology and medicine'.[10] No Korean embassy went to Peking during these centuries without asking for the latest books on astronomy and mathematics, on geography and medicine. The envoys also asked for samples of instruments too and, as we shall see in Chapter 5, in Jesuit times at least they got them liberally.

The most widely held stereotype of the Kingdom of Chosŏn is that, after the days of its first few great monarchs, it wallowed in stagnation brought about by a dry Neo-Confucian orthodoxy and bureaucratic factionalism. That stereotype is in fact false, the product of an imperialist mentality, seen in the self-serving writings of Western missionaries and Japanese colonialists.[11] The Yi Dynasty was indeed committed to a Neo-Confucian ideology; and Confucianism has sometimes been seen as the villain in the 'failure' of China and the countries in its cultural orbit to experience a scientific revolution. Yet the activities that we describe in this book contradict the notion of Neo-Confucian 'intellectual stagnation' in Korea; and there is no evidence whatever to suggest that King Sejong's instrument-builders and King Hyŏnjong's horologists were any less staunch in their orthodoxy than the rulers they served. Moreover, to see Neo-Confucianism as necessarily an enemy of science is a bizarre idea in itself, for many have concluded that its world-view was very congruent with that of modern science,

[10] Needham, SCC III: 298, 302, 389–90, 431, and a special appendix devoted to Korea, pp. 682–3; SCC IV.2: 516–22, esp. p. 519; SCC IV.3: 453; SCC V.2: 201; SCC V.3: 167, 177.

[11] We are grateful to Dr Gari K. Ledyard and to Dr Nathan Sivin for useful discussions on this point.

much more so indeed than that of traditional Western theology and philosophy.[12] To have a glimpse of evolution, and to build one's whole universe out of *ch'i* 氣, matter-energy, and *li* 理, pattern-principles at all levels, betokened remarkably enlightened minds.[13]

Akin, though opposite in intent, to Hulbert's anti-Confucian bias is the view that Confucianism and science were antithetical in another sense: that such things as the instruments we describe herein were triumphs of a Confucian society, but that they were not science. King Sejong, in this view, cannot be described as a patron of science despite his having spent a fortune on, and taken a strong personal interest in, the construction of new astronomical instruments. Rather, he was simply a more than usually competent monarch doing his job; that job included having an observatory, so he built one. Or again, in this view, King Sejong's rain-gauges had nothing to do with science, because they were part of a larger scheme for reforming the basis of land-tax assessment. The perceptive idea that measuring precipitation could give valuable information about the differential productivity of agricultural land was, it would thus be maintained, not 'scientific', but merely an example of administrative intelligence.

We take note of this view only in order to reject it firmly. It seems to us to arise from a fundamentally false historical perspective; yet here, in the history of science, a correct one is truly a *sine qua non*. In our view China, Korea, India, and the Arabs all had science, as also the ancient and medieval Europeans, and all made valuable contributions to it. But *modern*, ecumenical, and universal science, in other words the science of the Scientific Revolution, originated only in Europe during the late Renaissance. It is useless to look for hydrodynamics, electronics, or organic chemistry in ancient and medieval civilisations, or to complain of their absence, for these things are characteristic of modern science. During the past three centuries this has spread throughout the world, and there is no one, of whatever race, sex, colour, or creed, who cannot use it and add to it if once trained in it. Modern science, based upon the mathematisation of hypotheses about Nature, and upon relentless experimentation, has shed the ethnic limitations

[12] Needham, SCC II: 472 ff, 493 ff.

[13] This is borne out by Yung-sik Kim in his 'The World-View of Chu Hsi (1130 to 1200): Knowledge about the Natural World in the *Chu Tzu Ch'üan Shu*' (unpub. doctoral thesis, Princeton University, 1979), though Kim is critical enough of the details of Neo-Confucian proto-scientific observations and conclusions. Also relevant to this point is the excellent doctoral dissertation of Park Song-rae, 'Portents and Politics in Early Yi Korea, 1392–1519' (unpub. doctoral thesis. University of Hawaii, 1977).

which characterised the ancient and medieval sciences of various cultures; but these are not, as some have believed,incommensurable and eternally incompatible, like forms of artistic creation – they all embodied real advances in human knowledge; they were part of one single great movement in social evolution. Ever since man became able to observe, to reason, and to record, Nature has been much the same, so that all advances in natural knowledge, however haltingly the theories about them were expressed, have been real and durable. It was only in Galileo's time that the best method of discovery was itself discovered. In other words, we visualise the sciences of all the traditional ancient and medieval civilisations flowing into the ocean of modern science like rivers to the sea.

It would thus be culture-bound and deeply unfair to deny the name of science to the many scientific traditions which existed before our own modern science of the post-Renaissance West. The systematic application of human intelligence for the acquisition of knowledge about the natural world (whether for its own sake or in order to accomplish tasks through the development of technology) has been a universal human activity, and that activity is science. The Korea of the Yi Dynasty was profoundly Confucian, culturally proud, dynamic, and engaged in the science and technology of astronomy and horology in ways that ought to capture the imagination of all who are interested in the history of mankind's continuing discovery of the world of Nature.

In order to thank the many friends on whom we have relied for assistance and advice, we should like to describe briefly how this book came to be written.

In the 1950s, in the course of research which was to result in the book *Heavenly Clockwork*,[14] Joseph Needham, Wang Ling, and the late Derek J. de Solla Price became aware of a Korean demonstrational armillary sphere that had first been described in Western literature by Rufus and Lee in 1936.[15] Rufus's photograph of that instrument was reproduced as Fig. 59 of *Heavenly Clockwork*, and again as Fig. 179 in volume III of *Science and Civilisation in China*.

During the Japanese occupation of Korea, the instrument had been in the collection of Mr Kim Sŏngsu, who later presented it to the Koryŏ University

[14] (Cambridge, Cambridge University Press, 1960; hereafter Needham, Wang, and Price, HC.)

[15] W. Carl Rufus and Lee Won-chul, 'Marking Time in Korea', *Popular Astronomy*, 1936, *44*: 252–7; also Rufus, 'Astronomy in Korea', *Transactions of the Korea Branch of the Royal Asiatic Society*, 1936, *26*: 1–52, pp. 38–9 and Fig. 26.

Museum in Seoul.[16] Further enquiries showed that the instrument had fortunately survived the Korean War of 1950–3. Derek Price, with the co-operation of Dr Silvio Bedini, attempted to obtain the approval of the Korean authorities for the transport of the instrument to the Smithsonian Institution for a period of study. It was not possible to arrange for that to be done, but the enquiries had the good effect of stimulating the interest of the Korean government in the instrument, with the result that it was soon registered as a National Treasure.

Early in 1962, John Combridge's collaboration with Joseph Needham in the further study of East Asian clockwork mechanisms resulted in the translation, in draft, by Needham and Lu Gwei-djen, of chapters from the *Sejong sillok* and the *Chŭngbo munhŏn pigo* relating to scientific instruments. Study of these passages led, *inter alia*, to the belief that the instrument in the Koryŏ University Museum might be that constructed in 1669 by Song Iyŏng, incorporating an armillary sphere made by Yi Minch'ŏl. The importance of the identification of that instrument, if the hypothesis were to prove correct, prompted further efforts at investigation.

In 1963 the aid of Professor Gari K. Ledyard, then resident in Seoul, was enlisted in obtaining photographs of the demonstrational armillary sphere and its clockwork mechanism. Through the good offices of Dr Jeon Sang-woon of the Sungshin Women's Teacher's College, Seoul, and with the kind co-operation of Professor Kim Chŏnghak, Director, and Mr Yun Seyŏng, Curator of Historical Collections of the Koryŏ University Museum, Ledyard was able, the following year, to supply Needham and his colleagues with an excellent set of thirty detailed photographs. Several similar photographs were subsequently published for the first time by Jeon in *Science and Technology in Korea* (1974).[17]

The photographs became the basis for a detailed study of the instrument by Combridge, which was completed in draft in 1964.[18] That study reinforced the belief that the instrument was indeed that of Song Iyŏng and Yi Minch'ŏl. It was

[16] Kim Sŏngsu, a noted antique collector, was an important industrialist, the founder of Korea's leading newspaper, and the builder of several educational institutions including Koryŏ University. After World War II he was active in Nationalist politics and was Vice-President of the Republic of Korea during 1950–1. (We owe this information to Gari K. Ledyard, private comm.)

[17] Jeon, STK, pp. 68–72 and Figs. 1.17, 3.11, 3.12, and 5.6. See also Jeon, 'Senki gyokkō (tenmon tokei) ni tsuite' 璇璣玉衡(天文時計)について (On armillary spheres with clockwork in the Yi Dynasty of Korea), *Kagakushi kenkyū* 科學史研究, 1962, *63*: 137–41; 'Yissi Chosŏn ŭi sige chejak sogo' 李氏朝鮮의時計製作小考 (A study of timekeeping instruments in the Yi Dynasty), *Hyangt'o sŏul* 鄕土서울, 1963, *17*: 49–114, pp. 102–11.

[18] (T. O. Robinson), 'A Korean 17th Century Armillary Clock' (notes on a lecture by J. H. Combridge on 27 November 1964), *Antiquarian Horology*, March 1965, *4*: 300–1.

then decided that the study was of such significance that it should become the basis for a monograph on the astronomical instruments of the early and middle Yi period which would also incorporate the passages previously translated from the *Sejong sillok* and the *Chŭngbo munhŏn pigo*.

There matters rested for several years. In 1973, Joseph Needham asked John Major if he would undertake the task of bringing the monograph to completion. Working together over the next few years, Major and Combridge revised the technical study of the Song Iyŏng / Yi Minch'ŏl instrument. Meanwhile, in July of 1974 Major visited Seoul, and again through the good offices of Jeon was kindly permitted by the authorities of the Koryŏ University Museum to examine the instrument closely. This allowed the answering of some technical questions that could not be resolved through an examination of the photographs. At about the same time, the authors agreed that the photographs selected to accompany the study would be easier to understand if they were augmented by a set of explanatory line-drawings.

Meanwhile, work proceeded on the revision of the earlier draft translations from the *Sejong sillok* and the *Chŭngbo munhŏn pigo*; and photographs of surviving parts of King Sejong's instruments, and of other instruments analogous to them, were assembled. As the technical details contained in the translations were understood with greater clarity, drawings and sketches of the instruments were produced.

With the revision of an article by Needham and Lu on an eighteenth-century astronomical screen, previously published in 1966 in *Physis*,[19] and with the writing of additional chapters to set our research in an appropriate historical framework, the work was brought to completion.

Throughout this long span of time, Gari K. Ledyard offered considerable further assistance, making numerous corrections to, and comments on, the draft translations and providing us with a detailed critique of a draft of the entire book. Similarly, Jeon Sang-woon freely supplied advice on questions of Korean history and technology, and furnished us with a number of the photographs used herein.

We wish to express our deep gratitude to Dr Ledyard, Dr Jeon, and the other colleagues and friends named above, without whose aid this work would never have been accomplished. In addition, thanks are due to Dr Nathan Sivin for much

[19] 'A Korean Astronomical Screen of the Mid-Eighteenth Century from the Royal Palace of the Yi Dynasty (Chosŏn Kingdom, 1392 to 1910)', *Physis*, 1966, *8.2*: 137–62.

helpful advice during the later stages of the work, to Dr David C. Major, who read a complete draft of the manuscript and made many helpful suggestions, and to Mr William Taylor for advice on drawings.

We are grateful to Mrs Katherine Lowe, Mrs Elizabeth Alexander, and Mrs Gail Patten for their patient and accurate work on the many drafts of the manuscript that have passed through their hands.

Our most grateful thanks are due to Mr Philip Robinson of London for a grant of £1000 towards the illustrations and the cost of production of this book. We are also very grateful to Mr David Penney F.B.H.I. for the excellent drawings which he has made. Finally, we acknowledge with thanks the financial support of the Dartmouth College Committee on Faculty Research; and that of the Wellcome Trust of London, which provided funding for Lu Gwei-djen during the time when this book was being written; as also the facilities and financial support of the East Asian History of Science Trust and Library at Cambridge. John Major expresses his gratitude to the President and Members of Clare Hall, Cambridge, where he was a Visiting Fellow when this book was completed.

JOSEPH NEEDHAM
LU GWEI-DJEN
JOHN H. COMBRIDGE
JOHN S. MAJOR

1

THEORETICAL AND HISTORICAL BACKGROUND

FIRST CONSIDERATIONS: KINGSHIP AND COSMOLOGY

In the Chinese tradition shared by Korea, astronomy was both a royal duty and a royal prerogative. This had been true from time immemorial, and the Chinese ascribed the invention of astronomy to the legendary sage-emperors Yao 堯 and Shun 舜. In classical Chinese cosmology, the emperor occupied a place at the axis of the universe, mediating between Heaven and Earth. He 'sat on his throne and faced south'; ritually, if not in fact, his throne was directly beneath the pole star, and, facing south, he surveyed the whole world. The emperor had a sacred duty to promulgate an accurate calendar so that his actions could be in accord with the movements of the heavenly bodies; and he employed astronomers – who were also portent astrologers – to watch for celestial anomalies that might require an imperial response.

The rulers of kingdoms peripheral to China usually accepted this theory, and while they might acknowledge the celestial and temporal primacy of the emperor of China – the kings of Korea usually did so – they also placed themselves firmly at the *axis mundi* of their own smaller realms. Thus the Royal Observatory of the Korean Kingdom of Silla 新羅, the Ch'ŏmsŏngdae (Chan-hsing T'ai 瞻星臺, 'The Estrade for Gazing at the Stars'), built in 647 (Fig. 1.1), was also called the Pidu (Pi-tou 比斗), 'Comparable to the (Northern) Dipper'.[1] Similarly, the terrace on

[1] Jeon, STK, p. 35. See also Song Sang-yong, 'A Brief History of the Study of the Chŏmsŏng-dae in Kyŏngju', *Korea Journal*, Aug. 1983, *23*.8: 16–21. The term Pidu admits of several possible interpretations. It could mean 'Comparable to the (Northern) Dipper', but also '(The Place for) Comparing the Dipper', i.e. observing the directional change of the 'handle' or 'pointer' of the Dipper at any given hour of the night throughout the year (its nightly directional shift is one celestial degree, i.e. $\frac{1}{365.25}$ of the celestial circle; see n. 14 below). One might also suspect here an early conflation of the similar words *pidu* 比斗 and *paedu* (*pei-tou* 北斗), 'Northern Dipper', as a name for the observational tower in popular parlance.

The 'Dipper' or 'Northern Dipper', is the well-known group of seven bright stars which are the most conspicuous part of the constellation Ursa Major. In the United States it is usually called (by coincident pictorialisation) the Big Dipper; in England it is often loosely termed the Great Bear, but the less common term the Plough is more accurate, since this refers to those seven stars alone as distinct from the entire constellation.

which stood the Japanese royal palace during the Nara period (eighth century) was known as Hokudodai (Pei-tou T'ai 北斗臺), 'Northern Dipper Terrace'.[2] These names show that the early Korean and Japanese kings accepted the Chinese cosmological principle of a connection between royal authority and celestial polar centrality. That connection being understood, it was deemed essential for any king who ruled in the Chinese tradition to engage in astronomy and calendar-making, and to employ court astronomers for that purpose. The Royal Observatory of the Yi Dynasty, the Sŏun Kwan, was charged with carrying out these astronomical, calendrical, and meteorological duties. Nevertheless, the fact that the Sŏun Kwan, and other East Asian royal observatories like it, owed their royal patronage in part to a cosmologically based theory of kingship does not mean that their methods and achievements were unscientific.[3]

THEORETICAL BACKGROUND

Although it would clearly not be possible, in these few pages, to summarise adequately the whole history of Chinese astronomy down to the end of the fourteenth century, a history that forms the proper background to the astronomy of the Yi Dynasty, we nevertheless think it useful to present here, in outline form, a few key concepts relating to cosmology, calendrical science, horology, and astronomical instruments. The reader may wish to consult in addition the list of definitions of important technical terms in astronomy to be found in the third volume of *Science and Civilisation in China*.[4]

A. Cosmology.

1. Cosmological theories. While a variety of cosmological theories (particularly the *kai-t'ien* 蓋天, 'Canopy Heaven'[5] theory) had competed for primacy in the early days of Chinese astronomy, after the Han period the *hun-t'ien* 渾天 ('Enveloping Heaven') theory was generally accepted as

[2] We are grateful to Professor Fujieda Akira of the Research Institute for Humanistic Studies, Kyoto University, for this information. Professor Fujieda conducted one of us (JSM) on a tour of the archaeological excavations of the site in 1975.

[3] N. Sivin, *Cosmos and Computation in Early Chinese Mathematical Astronomy* (Leiden: E. J. Brill, 1969).

[4] Needham, SCC III: 178–82.

[5] We employ this term as a translation for *kai-t'ien* in preference to 'Heavenly Cover', as used in Needham, SCC III. The concept involved is that heaven is shaped like the domed lid (*kai*) of an ancient Chinese bronze or ceramic vessel, or like the domed parasol canopy (also *kai*) of a Chinese chariot, covering the flat earth. The term is grammatically parallel to *hun-t'ien*, 'Enveloping Heaven'.

Fig. 1.1. The Ch'ŏmsŏngdae, the observatory of the Kingdom of Silla, Kyŏngju, 647 C.E. The tower was presumably used for celestial observations, and may have had instruments mounted on the platform at its apex. The tower itself may also have served as a gnomon. The square structure atop the tower has a true N–S and E–W alignment.

correct.[6] This theory assumed that the universe was spherical and centred on a flat earth. The sphericity of the earth, despite early hints in that direction, was a relatively late refinement of this theory.[7] All armillary spheres in the Chinese tradition model the *hun-t'ien* universe.

2. Equatorial-polar coordinates. From its earliest development, Chinese astronomy was equatorially based.[8] This stands in marked contrast to the ecliptic-polar basis of Western astronomy prior to Tycho Brahe. From a computational point of view there is little to choose between the two coordinate systems,[9] but the Chinese equatorial system had important consequences for the development of astronomical instruments, as we shall see below, since the mounting of armillary spheres in the Chinese tradition reflected it.
3. Divisions of the celestial sphere. For cosmological and observational purposes, the visible part of the celestial sphere was divided, in the Chinese tradition, in several ways (Fig. 1.2):[10]
   a. The five palaces. One of the simplest, and one of the oldest, divisions of the heavens was into five 'palaces'. These comprised the central circle of the north circumpolar stars (which never, for an observer in northern China, dip below the horizon), called the *tzu-wei kung* 紫微宮, 'Palace of Purple Tenuity'; and four truncated sectors extending from the circle bounding the north circumpolar stars to the south circumpolar circle of perpetual invisibility, these sectors being designated as the palaces of the East, South, West, and North. The five palaces are correlated with the Five Phases[11] (*wu hsing* 五行), the four cardinal directions (plus the centre), the seasons, etc. The four non-central palaces were symbolised by four animal emblems: the Blue-Green Dragon of the East, the Vermillion Bird of the South, the White Tiger of the West, and the Dark Warrior of the North (a paired turtle and snake).

[6] For convenient explanations of these theories see Shigeru Nakayama, *A History of Japanese Astronomy: Chinese Background and Western Impact* (Cambridge, Mass.: Harvard University Press, 1969), pp. 24–43.

[7] Needham, SCC III: 216–17; but see also Nakayama, p. 39; and Christopher Cullen, 'Joseph Needham on Chinese Astronomy', *Past and Present*, 1980, *87:* 39–53, p. 42.

[8] Needham, SCC III: 231.

[9] Cullen, p. 45.

[10] Léopold de Saussure, *Les origines de l'astronomie chinoise* (Paris, 1930), *passim.*

[11] For correlations of the Five Phases (also known as the 'five elements') see the table in Needham, SCC III: 262–3. See also John S. Major, 'Myth, Cosmology, and the Origins of Chinese Science', *Journal of Chinese Philosophy*, 1978, 5: 1–20, pp. 11–15.

Fig. 1.2. A bronze mirror of the Middle Koryŏ period, symbolising the various divisions of the northern celestial hemisphere. From the centre: a knob representing the *axis mundi* at the north pole; the four directional animal symbols; the eight trigrams; the twelve animal symbols of the Earthly Branches, representing the Jupiter Stations; the twenty-eight Lunar Lodges and the twenty-four Fortnightly Periods. Compare the similar Chinese mirror shown in Needham, SCC III: Fig. 93.

b. The nine fields. This was similar in conception to the five palaces, except that the portion of the sky surrounding the north circumpolar stars was divided into eight, rather than four, truncated sectors. The nine fields corresponded to the 'nine continents' into which the earth was schematically divided; the eight non-central fields were correlated with the eight trigrams of the *I ching* 易經 (Book of Changes).

c. Jupiter Stations. The equator (and, by analogy, the ecliptic) was divided into twelve 'Jupiter Stations', reflecting the ancient idea that the (approximately) twelve-year orbital period of Jupiter (and, more specifically, of *t'ai-sui* 太歲, an invisible counter-orbital correlate of Jupiter) was of great astrological (or, more properly, chronomatic) significance. The twelve Jupiter Stations were correlated with the twelve months and with the twelve 'Earthly Branches' (*chih* 支) of the sexagenary system for enumerating days and years. The Earthly Branches were in turn correlated with symbolical animals; the Jupiter Stations did not, however, equate to the twelve Signs of the Western zodiac.[12]

d. Lunar Lodges. The equator was also divided into twenty-eight unequal sectors called Lunar Lodges (*hsiu* 宿), defined by twenty-eight constellations lying mostly near an ancient celestial equator. This too was a system of great antiquity,[13] related to the similar Indian *nakshatras* (which also played a significant role in Islamic astronomy). The Lunar Lodges were derived from the $27\frac{1}{3}$ days of the moon's sidereal period, but anciently they were also commonly correlated with the $29\frac{1}{2}$-year orbital period of Saturn, approximated as 28 years. The location of celestial bodies within the Lunar Lodges was an important consideration in East Asian portent astrology.

[12] The term 'Chinese Zodiac' is the result of a misconception. See J. H. Combridge, 'Chinese Sexagenary Calendar-Cycles', *Antiquarian Horology*, Sept. 1966, 5.4: 134; and also 'Hour Systems in China and Japan', *Bulletin of the National Association of Watch and Clock Collectors, Inc.*, Aug. 1976, *18*.4: 336–8.

[13] The earliest extant list of all twenty-eight *hsiu* is found on the lid of a lacquer box, dated approximately 433 B.C.E., recently excavated in Hupei, China, and now in the Hupei Provincial Museum, Wuhan. See Wang Chien-min 王建民 *et al.*, 'Tseng Hou-i mu ch'u-t'u ti erh-shih-pa hsiu ch'ing-lung pai-hu t'u-hsiang' 曾侯乙墓出土的二十八宿青龙白虎图象 (On a picture of the twenty-eight Lunar Lodges, the Blue-Green Dragon, and the White Tiger, excavated from the tomb of the Marquis Yi of Tseng), *Wen-wu* 文物, 1979.7: 40–5. See also Needham, SCC III: 231–59. Some of the extensive literature in Chinese on the origins of the *hsiu* is cited in Xi Zezong, 'Chinese Studies in the History of Astronomy, 1949–1979', *Isis*, Sept. 1981, *72*: 456–70, p. 463.

e. Fortnightly Periods. The tropical year was divided into twelve equal periods, *ch'i* 氣, defined by 30° of solar motion along the ecliptic. These were subdivided into twelve 'mid-point *ch'i*' (*chung-ch'i* 中氣) and twelve 'nodal *ch'i*' (*chieh-ch'i* 節氣), making a total of twenty-four Fortnightly Periods nominally of 15 days but on average 15.219 days each. These periods were in effect a series of sub-seasons that defined an agrarian solar calendar for everyday use. In practice the Fortnightly Periods were counted in whole days, with extra days inserted as necessary to account for accumulated fractional days.

f. Degrees. Finally, the equator, the ecliptic, and all other celestial circles were divided into $365\frac{1}{4}$ 'degrees' (*tu* 度).[14] The replacement of this number by the 360 degrees of a circle as reckoned in the West was a consequence of the introduction of Jesuit methods into Chinese astronomy.

B. The calendar.

1. Nature of the calendar. A Chinese calendar (*li* 曆) was not merely a device for keeping track of days and seasons, but rather a system of astronomical calculations that yielded a complete ephemeris of celestial motions. Its most important task, and most serious difficulty, lay (as Nakayama has noted) in 'reconciling two fundamentally incommensurable periods – the tropical year and the synodic month'.[15] The calendar had also to reconcile the tropical year with the sidereal year. A satisfactory calendar was expected also to predict solar and lunar eclipses with accuracy; and although Chinese calendrical science devoted relatively less attention to the orbital periods of the five visible planets, such matters as conjunctions, occlusions, oppositions, and other phenomena of planetary location within the Lunar Lodges and in relation to each other were important in portent astrology and demanded a certain amount of attention.[16]

2. Calendar reform. The magnitude and complexity of the task faced by the designers of Chinese calendars, and the difficulties of precision plotting of

[14] In order to prevent any ambiguity between Western degrees and the slightly smaller Chinese celestial degrees, throughout this book the former will be denoted by the conventional superscript °, and the latter by a superscript ^{d} (denoting both 'degrees' and 度 in its *pinyin* romanisation *du*). Thus, for example, a circle consists of 360° and $365\frac{1}{4}^{d}$.

[15] Nakayama, p. 67.

[16] *Ibid.* pp. 150–1.

celestial orbits in the absence of the telescope and, more important, an adequate theory of celestial mechanics, meant that calendars inevitably accumulated errors and had to be revised from time to time. The Bureau of Astronomy often tended to consult the calendar rather than the stars for reckoning celestial time, but when the differences between the two became too great, the making of a new calendar was one of the Bureau's most important tasks.[17]

3. The political significance of calendar reform. The proclamation of a new calendar was an essential symbol of the assumption of imperial authority by each new Chinese dynasty. The 'new' calendar might be simply the previous dynasty's calendar under a new name, with or without minor revisions, or it might be a truly new and reformed calendar. New calendars were sometimes also promulgated, with appropriate rituals, during the lifetime of a dynasty. The kings of Korea, as tributary vassals of the Chinese emperor, had no choice but to accept and respond to these new calendars as they appeared.

C. Timekeeping. The Chinese day began at midnight. The Hebrew/Greek custom of beginning the day at sunset was unknown in the Chinese tradition. The Chinese day was divided in several ways:[18]

1. Twelve 'double-hours' (*shih* 時), each corresponding to two hours of Western time. The double-hours were divided into halves, named for their 'beginnings' (*ch'u* 初) and 'mid-points' (*cheng* 正). Their timing was such that midnight and noon came at the 'mid-points' rather than at the 'beginnings' of the corresponding double-hours; that is, the double-hours ran from 11 p.m. to 1 a.m., 1 a.m. to 3 a.m., etc.
2. One hundred equal intervals called *k'o* 刻, each corresponding to 14 min 24 sec of modern (Western-derived) time-reckoning. The *k'o* were subdivided into fractions, called *fen* 分. Because each double-hour corresponded to $8\frac{1}{3}$ *k'o* and each half-double-hour to $4\frac{1}{6}$ *k'o*, the number of *fen* was usually six or some multiple thereof. With the adoption of Western time-reckoning in China in the seventeenth century, the name *k'o* passed

[17] *Ibid.* p. 67.

[18] Jeon, STK, pp. 87–93; Needham, Wang, and Price, HC, pp. 199–205.

into general use for integral quarter-hours, each of 15 Western minutes (to the latter of which the word *fen* was applied).[19]

3. Night-watches. In addition to the two forms of constant 'clock-time' just mentioned, the Chinese divided each nominal night, between the officially defined times of 'dusk' and 'dawn' (or, anciently, between sunset and sunrise), into five equal but seasonally variable 'night-watches' (*keng* 更), each of which was divided into five equal portions (*tien* 點 or *ch'ou* 籌). Because of the cyclic variation in the length of the night over the course of a year, the automatic striking (on drums and gongs) of the night-watches and their divisions, and the moments of 'dusk' and 'dawn' as well, presented a formidable (and elegantly overcome) problem to the designers of timekeeping instruments in the Sino-Korean tradition.[20]

D. Instruments. As we have noted, Chinese astronomy was equatorial rather than ecliptic in orientation; it was also arithmetical-algebraic rather than geometrical in its methods. Chinese astronomers showed little interest in formulating intellectual models of the cosmos of the sort that dominated Western astronomy from Aristotle to Copernicus and Tycho Brahe and beyond. Whether this was, on the whole, an advantage or a disadvantage (or neither) for Chinese astronomy in the long run is a matter of debate; it did, however, have important consequences for astronomical instrumentation in the Chinese tradition.

1. Chinese instruments, from the Han 'diviner's board' (*shih* 式)[21] and the Later Han armillary spheres of Chang Heng 張衡 onwards, were equatorial in their reference.[22]
2. The armillary sphere, along with the gnomon, was the fundamental

[19] To avoid ambiguity as to the actual time-spans involved, in this book we confine the translations 'quarter-hour' and 'minute' to instances where the modern senses are appropriate. Where the older senses of *k'o* and *fen* are concerned, we employ the translations 'interval' and 'fraction', adding the Chinese words in romanisation in parentheses where necessary to indicate that the terms are used in a technical sense to denote specific periods of time. The word *k'o* means literally a 'notch' or 'mark' on a timekeeping scale; our term 'interval' is not, strictly speaking, a translation, but rather is a paraphrase intended to emphasise that the word denoted a precise interval of time.

[20] Needham, Wang, and Price, HC, pp. 37–9; John H. Combridge, 'The Astronomical Clocktowers of Chang Ssu-hsun and his Successors, A.D. 976 to 1126', *Antiquarian Horology*, June 1975, *9.3*: 288–301, p. 293; Jeon, STK, pp. 88–93; Needham, SCC IV.2: 517–18.

[21] Donald Harper, 'The Han Cosmic Board (*shih* 式)', *Early China*, 1978–9, *4*: 1–10. See also Christopher Cullen, 'Some Further Points on the *Shih*', and Donald J. Harper, 'The Han Cosmic Board: A Response to Christopher Cullen', *Early China*, 1980–1, *6*: 31–46, 47–56.

[22] Needham, SCC III: 342–54 *et seq.*

instrument of Chinese-style observatories. The equatorial basis led to a variety of other instruments as well, such as the equatorial sundial and Kuo Shou-ching's 'equatorial torquetum', the Simplified Instrument (*chien-i* 簡儀).

3. The equatorial basis suggested the desirability and mechanical feasibility of armillary spheres (*hun-i* 渾儀) and celestial globes (*hun-hsiang* 渾象) that were rotated automatically.[23]
4. The automatic rotation of instruments began, so far as surviving records show, in the time of Chang Heng (*c.* 117 C.E.) and reached a high point with the great astronomical clocktowers made by Chang Ssu-hsün 張思訓 and his successors, including Su Sung 蘇頌 and associates, during the Sung period. These clocktowers were powered by great 'timekeeping waterwheels',[24] the operating principles of which were based conceptually on those of the steelyard clepsydra.[25] These wheels found no direct descendants among the automatic timekeepers of Korea, however. The latter were based instead on other types of clepsydral technology[26] imported from Central Asia into China during the Yüan period.

We conclude this introductory section by noting that the immediate influences on the astronomy and instrumentation of the Korean Royal Observatory at the beginning of the Yi period were Kuo Shou-ching's Shou-shih 授時 'calendar' (more accurately, a complete system of astronomical calculations) of 1280,[27] his full set of instruments for the Chinese Imperial Bureau of Astronomy of about the

[23] *Ibid.* 359–66.

[24] Needham, Wang, and Price, HC, *passim*; John H. Combridge, 'The Celestial Balance', *Horological Journal*, Feb. 1962, *104*.2: 82–6; Needham, SCC IV.2: Fig. 658; Combridge, 'Astronomical Clocktowers'; Combridge, 'Clocktower Millenary Reflections', *Antiquarian Horology*, Winter 1979, *11*.6: 604–8.

[25] Needham, Wang, and Price, HC, p. 57, n. 2, para. 4; p. 94.

[26] Donald R. Hill, ed. and tr., *On the Construction of Water-Clocks: Kitāb Arshimīdas fī 'amal al-binkamāt* (London: Turner and Devereux, Occasional Paper No. 4, 1976); Hill, *Arabic Water-Clocks* (Aleppo, Syria: University of Aleppo Institute for the History of Arabic Science, 1981); Needham, Wang, and Price, HC, pp. 88, 97, 121ff, 140, 163, 186–7, and Fig. 34.

When we began this study we had expected to find that at least some of the mechanical timekeepers of Korea during the Yi period were directly descended from the Sung timekeeping waterwheel clocktowers of Chang Ssu-hsün *et al.* (see n. 24 above), and we were surprised to find that such is not the case as far as the driving mechanism is concerned. No doubt the Sung instruments stimulated the *idea* of mechanical timekeeping in Korea as well as in post-Sung China. But the Korean instruments, as we shall see in Chapter 2, belong primarily to a Sino-Arabic clepsydral tradition; only the time-annunciating jackwork of the Sung clocktowers was reflected in their design.

[27] Nakayama, pp. 123–50.

same time,[28] the time-annunciating jackwork (but not the timekeeping waterwheel technology) of the Sung astronomical clocktowers, and the Sino-Arabic tradition of striking clepsydras, with bells, drums, etc. and jackwork operated by falling metal balls.[29]

HISTORICAL BACKGROUND: FROM KORYŎ TO YI[30]

By the early thirteenth century, the Koryŏ Dynasty had been ruling Korea for three hundred years. For the latter half of that span of time, the Koryŏ kings had been much put upon by the 'barbarian' dynasties that had established themselves in Manchuria and North China in competition with the Chinese Sung Dynasty. Those experiences, however, paled into insignificance beside the invasions suffered by Korea at the hands of the new Mongol confederation of Genghiz Khan in the early decades of the thirteenth century. After almost thirty years of these invasions, the Koryŏ kings became vassals of the Mongol rulers of China (who were shortly to proclaim themselves the Yüan Dynasty) in 1260.

Kubilai Khan, the first Yüan emperor, proved to be a great patron of astronomy. His greatest Astronomer-Royal, Kuo Shou-ching, produced a magnificent set of new instruments for the Royal Observatory in the 1270s (Fig. 1.3), and his Shou-shih astronomical system was promulgated in 1280. The Koryŏ kings accepted the new calendar as a matter of course,[31] and were well aware of the fame of the new instruments. The impact of these great events in Chinese astronomy was limited in Korea, however; for various reasons the Koryŏ kings were in no position to take significant initiatives in astronomy and instrumentation in response to them.

Korea had been severely weakened economically by the long years of the Mongol invasions, and the royal purse was thin. The power of the Koryŏ kings was fragmented and hedged about by strict overlordship on the part of the Mongol emperors, powerful aristocratic domination of the Korean throne

[28] Needham, SCC III: 367–72; Alexander Wylie, 'The Mongol Astronomical Instruments in Peking', in his *Chinese Researches* (Shanghai, 1897; repr. Taipei, 1966), part III: 'Scientific' (separately paginated), 1–27.

[29] See n. 26 above.

[30] Unless otherwise noted, information in this section is derived from Hatada Takashi, *A History of Korea*, tr. W. W. Smith, Jr, and B. H. Hazard (Santa Barbara: University of California (Santa Barbara) Press, 1969), pp. 51–66; and from William E. Henthorn, *A History of Korea* (New York: Free Press, 1971), pp. 117–23, 128–55.

[31] Jeon, STK, pp. 78–9.

coupled with crippling aristocratic factional strife, and the political and economic power of Buddhist monasteries. The country was weakened even further by being forced to contribute heavily to the ill-fated Mongol attempts to invade Japan in 1274 and 1281. These unsuccessful invasions hurt Korea far more than they hurt the Mongols themselves. In their aftermath, Yüan domination of the Koryŏ kings became even more complete. The Koryŏ kings were usually sons-in-law of the Mongol emperors thereafter, and were limited in their ability to exercise the usual royal prerogatives – such as the practice of astronomy.

From the middle of the fourteenth century, Mongol power itself began to weaken, and with it the effectiveness of the military power available to the Koryŏ kingdom. Throughout the remainder of the century, Japanese pirates (*Wakō* 倭寇) made increasingly severe raids along the Korean coast.

The fall of the Yüan Dynasty to the new Ming Dynasty in 1368 left the Koryŏ Dynasty in an isolated and vulnerable position. There was much vacillation at the Koryŏ court about whether to capitulate to the Ming (and accept the new Ming Ta-t'ung 大統 calendar, a slightly revised version of the Shou-shih system) or to remain loyal to the fallen Yüan.[32] This indecisive attitude aroused the suspicion of the Ming court; suspicion turned to enmity when the Koryŏ kings decided, in the end, to remain loyal to their old overlords. The next-to-last Koryŏ king decided in 1388 to act on his loyalty to the Yüan by sending General Yi Sŏnggye 李成桂, who had won great fame for his victories over the *Wakō* raiders, with an expeditionary force to invade Ming China. General Yi led his army to the Yalu River but, being convinced of the futility of the anti-Ming policy, refused to cross it; he brought his troops back to the capital at Kaesŏng and seized power from the king. Yi Sŏnggye briefly installed a new king, the last of the Koryŏ Dynasty, and then mounted the throne himself as the founder of a new dynasty in 1392. He is known posthumously as T'aejo, the Grand Progenitor, of the Kingdom of Chosŏn.

T'aejo quickly sought to establish good relations with the Ming, although he was unable to effect a reconciliation until after the death, in 1399, of the Koreanophobe first Ming emperor. He initiated a sweeping (and partly successful) land

[32] The Koryŏ king did in fact accept the Ta-t'ung calendar in 1370, but Koryŏ continued to be loyal to the Yüan in most other respects. Dr Gari Ledyard has pointed out to us, however (private comm.), that this was a complicated political issue, and remained so throughout the period 1368 to 1390, when the Ming defeated the Mongols in Manchuria. There were both pro-Yüan and pro-Ming factions at the Koryŏ court after 1370. See also Jeon, STK, p. 79; and Hatada, pp. 59–60.

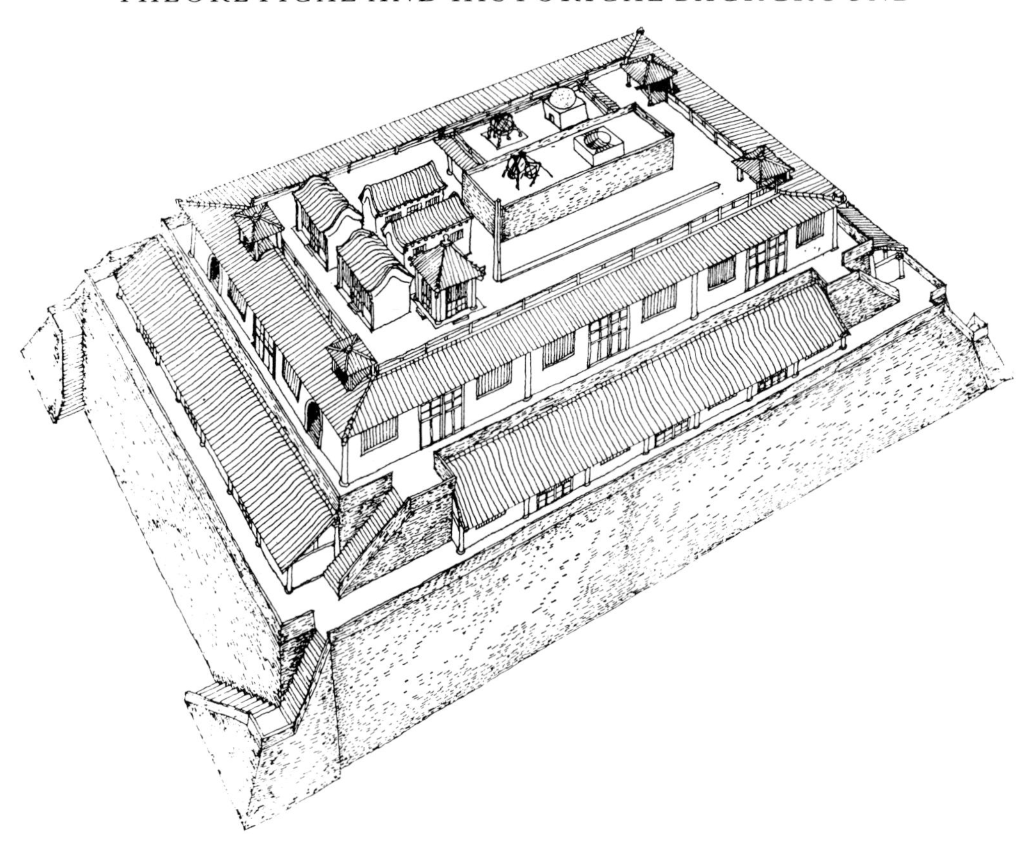

Fig. 1.3. Kuo Shou-ching's observatory at Peking, *c.* 1280: reconstruction drawing; view from the south-east. Visible on the northern part of the upper terrace are (clockwise from top): celestial globe, scaphe sundial, shadow-scale with 40-foot gnomon, Simplified Instrument, and armillary sphere.

reform which weakened aristocratic power, and he weakened it still further by promoting Chinese-style Confucian bureaucratism in place of aristocratic rule. He also patronised Neo-Confucianism as a state ideology, at the expense of Buddhism,[33] and broke the economic power of the monasteries by confiscating their extensive landholdings. The power of the throne was further enhanced by moving the capital to Seoul, away from the Buddhist and aristocratic establishments at Kaesŏng. T'aejo also continued the vigorous defence of the Korean coast against the Japanese pirates, on which he had built his reputation; he thus secured for his new dynasty external as well as internal peace. In the midst of these affairs

[33] Hatada, p. 63.

of state he also found time to pay attention to astronomical and horological matters; in 1395 he ordered the engraving of a new celestial planisphere (see Chapter 5 below) and in 1398 he ordered a new clepsydra to be set up in public in the capital.[34]

T'aejo bequeathed to his successors the foundations of a reformed and prosperous land economy, a respite from court factionalism, internal and external peace, and good relations with the Ming. He abdicated in favour of a son in 1398.

The second Yi king, after a brief reign, abdicated in favour of his younger brother in 1400. T'aejong, as the latter was posthumously to be known, immediately took advantage of the good conditions of the early Yi period to promote learning in all fields. In 1402 he laid the foundations for the dynasty's Bureau of Astronomy by ordering twelve scholars to make a study of astronomical matters.[35] In the next year he ordered the casting of a great fount of bronze moveable type and thereafter encouraged the printing of works of learning.[36] He patronised the Royal Academy and encouraged calendrical studies. He and his talented younger son, the future King Sejong, are said to have taken part personally in the construction of a new clepsydra.[37]

In 1418 King T'aejong abdicated in favour of his son (posthumously known as Sejong; r. 1418–50) and continued, in retirement, to protect the new king from the jealousy of his elder brothers. King T'aejong chose his successor well; Sejong's brilliant accomplishments have made him a Korean national hero.[38] King Sejong was a great statesman and a patron of all fields of learning. A notable part of the legacy that he bequeathed to his successors, as we shall see in the following chapter, was one of the finest astronomical observatories in the world.

To conclude this historical introduction, we note that while the fall of the Yüan Dynasty had powerful repercussions in Korea, including the fall of the Koryŏ Dynasty itself, the full flowering of the effects of those events in the field of astronomy had to await the third generation of Yi kings. The Koryŏ Dynasty was weak and nearly vestigial throughout the Yüan period, and was unable to respond effectively to the astronomical innovations of Kuo Shou-ching; the founder of the Chosŏn Kingdom was necessarily mainly concerned with dynastic consolidation.

[34] Jeon, STK, p. 56.
[35] Rufus, 'Astronomy in Korea', p. 29.
[36] Jeon, STK, pp. 175–7.
[37] Rufus, p. 29.
[38] Choi Hyon Pae, for the King Seijong [*sic*] Memorial Society, *King Seijong the Great* (Seoul, 1970).

By the reigns of T'aejong and, especially, Sejong, a combination of a highly favourable domestic socio-political and economic situation and the personal leadership of two talented and scientifically inclined kings made possible a renaissance of Korean science and technology. In the field of astronomy that renaissance was no doubt impelled all the more strongly by a perceived need, after a lag of more than a century, to equip the Korean Royal Observatory with instruments that would be as up to date technically as possible.

In Chapter 2 we turn to an examination of one of the most important results of the great outpouring of scientific energy under King Sejong: the re-equipping of the Royal Observatory.

2

THE RE-EQUIPPING OF THE ROYAL OBSERVATORY UNDER KING SEJONG

King Sejong (r. 1418–50), grandson of the founder of the Chosŏn Kingdom, led Korea during one of its greatest periods of cultural brilliance. His political achievements, though considerable, need not be discussed here; more important for the purposes of this study is his patronage of all fields of learning. Under his influence Neo-Confucianism rose to a position of eminence in Korea; first the court and then Korean society at large looked even more strongly than before to China as an intellectual model. One of Sejong's first acts as king was to re-establish the long-defunct Chiphyŏnjŏn (Chi-hsien-tien 集賢殿, 'The Hall of Assembled Worthies'), a sort of royal research institute modelled on a T'ang Dynasty academy of that name.[1] The Chiphyŏnjŏn grew rapidly in prestige; membership, limited to twenty scholars, was one of the greatest cultural honours that the dynasty could bestow. The academy was not merely a symbol of the king's patronage of learning; it played a direct role in many of the scientific achievements of Sejong's reign.

King Sejong himself, besides being a patron of culture and scholarship, had a personal talent for science and technology. He both ordered and had a direct hand in the making of improvements in printing technology, perfecting for the first time the use of moveable type,[2] and he commissioned the invention of the phonetic script known as Han'gŭl.[3] He sponsored improvements in military technology, including the design and construction of warships and new types of

[1] Henthorn, *A History of Korea*, pp. 140–1; Sohn Pow-key, Kim Chol-choon, and Hong Yi-sup, *The History of Korea* (Seoul: Korean National Commission for UNESCO, 1970), p. 132.

[2] Jeon, STK, pp. 177–81.

[3] Han'gŭl, originally known as *chŏngŭm* (*cheng-yin* 正音, 'correct sounds'), is one of the easiest and most rational phonetic scripts ever devised. Some authorities regard its invention as the single greatest achievement of King Sejong's reign. It made possible the creation of a Korean vernacular literature (although classical Chinese retained its prestige and official status as the language of learning and public affairs throughout the Yi period). See [Lee Sangbaek], 'The Origin of Korean Alphabet "Hangul"', (Part I of) Ministry of Culture and Information, Republic of Korea, *A History of Korean Alphabet and Moveable Types* (Seoul, n.d.), pp. 7–15.

firearms.[4] The king was particularly interested in astronomical matters; he ordered the correction of the Shou-shih and Ta-t'ung calendrical systems,[5] increased the staffing of the Bureau of Astronomy,[6] and undertook the complete re-equipping of the Royal Observatory.

The story of the great astronomical undertakings of King Sejong's reign is told in the *Sejong sillok* 世宗實錄 (Veritable Records of King Sejong) and is supplemented by accounts in the *Chŭngbo munhŏn pigo* 增補文獻備考 (Comprehensive Study of [Korean] Civilisation, Revised and Expanded). The chronicles preserve a number of inscriptions composed to be displayed alongside the instruments in the Royal Observatory. These, and especially the inscriptions and official memoirs of Kim Pin 金鑌 and Kim Ton 金墩, provide detailed descriptions of the instruments themselves, which we shall quote at length below. Their prefaces and summaries are couched in more general terms, and discuss the motives of the king and his officials in making the new instruments. While one must, in dealing with such documents, make allowances for the hyperbole of flattering courtiers, they nevertheless make it clear that the initiative of the king was a crucial factor in bringing about one of the most fruitful periods of astronomical instrument-making in East Asian history. Because they provide so much useful information about the context and motivation of this great project, we begin our account of King Sejong's instruments by quoting at length from some of the more general remarks contained in several of these inscriptions and memoirs.

Sejong sillok, 65: 1a, 2b–3a (1434 C.E.):

(On the first day of the eighth month, 1434), the new clepsydra was brought into use. The king had decided that the old clepsydra was not accurate enough, and had ordered that (metal parts for) a new one should be cast . . .

(The king) ordered Kim Pin to make an inscription for it, the preface of which said:

Among the policies of emperors and kings, none has been more important than the unification of times and seasons. The methods used for the study of these matters have been the armillary sphere, the celestial globe, the sundial, and the clepsydra. Without the sphere and the globe, there could be no study of the motions of the heavens and earth; without the sundial and clepsydra there could be no measure of the divisions between

[4] Jeon, STK, pp. 189–93, 212–13.

[5] *Chŭngbo munhŏn pigo* 1: 5b; Jeon, STK, pp. 79–80. It is not clear, however, how King Sejong could 'legally' do this, under the terms of Chosŏn's status as a tributary vassal state of China. Nor is it clear what 'corrections' were made.

[6] Jeon, STK, p. 105.

the days and nights. Over a thousand years, (at the correct moment) each one will start without any error. (This) can be achieved only if no neglect is permitted in the summation of the smallest differences in (gnomon) shadow length. Therefore all through time the sages have followed the heavens in their government; none has failed to respect this.

Now, His Majesty's servants, having in mind His profound respect for the Emperor Yao 堯, and imitating the example of the great Shun 舜 when he examined the (*hsüan-*) *chi* 璿璣 instrument[7] and ordered the artisans to make (the first) armillary sphere and celestial globe to verify and measure the several seasons, have constructed this new clepsydral apparatus in order to equalise the sundial and the intervals (*k'o*); and it is set up in the western part of the palace ...

As each hour comes round, the jackwork immortals (of the clepsydra) respond (with the appropriate time-signals). Consulting the (celestial) globe and the (armillary) sphere,[8] people find that (the time-signals) correspond to the movements of the heavens without the slightest mistake. It is really as if the gods and spirits were in charge of it. No one seeing it does not heave a sigh and aver that we Easterners [i.e. Koreans] certainly had nothing like this in former times ...

Having received the command, I, (Kim) Pin, have written this for future generations and humbly present the inscription, which says, 'Yin and yang follow each other, day and night alternately come round[9] ... The sundial and the clepsydra have long been made, but from the time of [the legendary sage-emperor] Huang-ti 黃帝 onwards there have been different methods, and only we Easterners have developed and extended the different designs ...'

Sejong sillok, 77: 7a (1437):

At the beginning of the year (1437), the king had ordered the construction of instruments to measure the days and nights ... The king commissioned Kim Ton to make an inscription, the preface of which said:

The making of celestial globes and armillary spheres is a high and ancient practice. From the emperors Yao and Shun down to the Han and T'ang (dynasties) there was no one who did not regard it as a most important thing. The literature about it is to be found in the classics and histories, but as we are far removed from ancient times, the methods

[7] See n. 11 below.

[8] This sentence, though somewhat misleadingly phrased, does not imply that a celestial globe and armillary sphere were incorporated into the Striking Clepsydra, but rather indicates reliance on those instruments to model the movement of the heavens. As we shall see below (Nos. 9 and 10, pp. 74–6), an armillary sphere and celestial globe mechanically rotated by means of a clepsydra were also produced as part of the re-equipping of the Royal Observatory.

[9] *Chiao-ts'o* 交錯. This phrase often appears in descriptions of the mechanisms of Chinese water-clocks; here it refers to the motions of the heaven themselves.

have not (been handed down) in great detail. Now His Majesty reverently took (the work of these) sages as the capstone of the achievements of antiquity. While resting from the myriad concerns of his duties, he turned his attention to the principles of astronomy and uranographic models. Accordingly, what were of old called armillary spheres, celestial globes, gnomons, Simplified Instruments, Automatic Striking Clepsydras, Small Simplified Instruments, hemispherical scaphe sundials, Horizontal Sundials and Plummet Sundials, all these instruments have been made without one missing. Such is His Majesty's respect for heavenly knowledge and for the exploitation of earthly things.

Sejong sillok, 77: 9b (1437):

It was ordered by the king that Kim Ton should write a memoir on the Simplified Instrument Observatory. This said:

In the seventh year of the (Ming) Hsüan-te 宣德 reign-period [= 1432] on a *jen-tzu* 壬子 day in the seventh month, His Majesty discussed the principles of calendrical and astronomical science (with his scholars and officials) in the Royal Lecture-Hall for the Exposition of the Classics. To the Academician Chŏng Inji 鄭麟趾 of the Academy of Arts and Letters, he said:

'We Easterners live far across the sea, but what we do is always based on the great cultural achievements of China; only in astronomical matters have we been somewhat deficient. You, My Minister, charged with the harmonisation of calendrical computations, shall, together with the Academician Chŏng Ch'o 鄭招, construct a Simplified Instrument for me.'

Sejong sillok, 77: 10b–11a (1438):

In the spring of the *wu-wu* 戊午 year (1438) the authorities concerned asked that a record should be made [of the previous several years' instrument-making activity] from beginning to end for the information of posterity. So they discussed their views with Your Servant [the historian], and I was ordered to write [the account that precedes this summary] about all of these affairs. As far as I can see, they were all based on [Kuo Shou-ching's] Shou-shih calendar [of 1280], which depends on the fundamental measurement of the heavens carried out by means of the armillary sphere and the celestial globe.

Therefore (as of old) when Yao ordered Hsi 羲 and Ho 和 to set in order the calendar by means of the sun and moon and the stars and constellations,[10] and when Shun

[10] *Shu ching* 書經 (The Book of Documents), 'Yao tien' 堯典 (The Canon of Yao). Tr. Bernhard Karlgren, 'The Book of Documents', *Bulletin of the Museum of Far Eastern Antiquities*, 1950, *22*: 1–81, p. 3; also James Legge, *The Chinese Classics* (Hong Kong and London, 1865), vol. III, part I, section 'The Shoo King', pp. 15–27, esp. pp. 18–22.

Hsi and Ho were, in Chinese mythology, originally a single individual. Hsi-ho was considered, in one tradition, the female charioteer of the sun-god; see David Hawkes, *Ch'u Tz'u: Songs of the South* (London: Oxford University Press, 1959), pp. 38, 49. In another tradition, Hsi-ho was the mother of the sun, or of ten

examined the (*hsüan-*)*chi* and the (*yü-*)*heng* 玉衡 to study the movements of the Seven Regulators [i.e. the sun, the moon, and the five visible planets],[11] so verily this is how to show sincere respect for the Heavens, which for the people's works and days cannot be neglected. From Han and T'ang onwards every dynasty has had its own instruments, some more accurate and some less; we cannot go into the details of all of them here. However, Kuo Shou-ching of the Yüan period, with his Simplified Instruments, his scaphe sundials, and his gnomons, etc., attained to the highest excellence of them all. Among what in former times we Easterners did, we have not heard of an instrument which demonstrated the march of the heavens as it revolved, but now this cultural achievement is also a reality.

Now His Majesty, in his sage wisdom and his profound respect (for Heaven), while resting from the myriad concerns of his duties, considered that the calendar was not as perfect as it ought to be; and ordered that it should be further studied, and better established; disturbed that measurements were not as accurate as they could be, ordered that (new) instruments should be constructed. How could Yao or Shun themselves have

suns; see Sarah Allen, 'Sons of Suns: Myth and Totemism in Early China', *Bulletin of the School of Oriental and African Studies*, 1981, *44*.2: 290–326, pp. 298–300. In the Chou period, when the mythical gods had begun to be regarded as 'sage-emperors' of high antiquity, Hsi and Ho were thought of as two court officials of 'emperor' Yao, in charge of astronomical matters. See Xi Zezong, 'Chinese Studies in the History of Astronomy, 1949–1979', p. 469.

[11] *Shu ching*, 'Shun tien' 舜典 (The Canon of Shun). Tr. Karlgren, 'Book of Documents', p. 5 (Karlgren treats the 'Shun tien' as a continuation of the 'Canon of Yao', and does not title it separately); Legge, *Chinese Classics*, III.1, 'Shoo King', pp. 29–51, esp. p. 33.

Chinese and other East Asian writers of treatises on astronomical instruments were endlessly fond of tracing the lineage of their efforts back to the legendary sage-emperors Yao and Shun. The inscription of King T'aejo's 1395 planisphere, translated in Chapter 5 below, pp. 158–9, provides another example of this.

Unfortunately, it is not possible to identify Shun's *hsüan-chi* and *yü-heng* 'instruments' with any certainty, nor to say just what he is supposed to have done with them. The seemingly comprehensive theory of H. Michel (on which see Needham, SCC III: 332–9) to the effect that the *hsüan-chi* was a notched jade disc used as a 'nocturnal' or 'circumpolar stellar template' for fixing the celestial north pole has now been shown to be false, at least with regard to the pre-Han period. Christopher Cullen and Ann S. L. Farrer present detailed evidence in support of this conclusion in 'On the Term *Hsüan Chi* and the Flanged Trilobate Jade Discs', *Bulletin of the School of Oriental and African Studies*, 1983, *46*: 52–76; see also Cullen, 'Joseph Needham on Chinese Astronomy', p. 46. The late Dr Derek Price, who was also preparing a paper on these jade discs, informed us (private comm.) that of the two dozen or so extant ones, about half are late fakes, while the rest may indeed have been used as 'nocturnals' or circumpolar stellar templates, as Michel described, though not in conjunction with the jade tubes (*tsung* 琮) that Michel identified with the *yü-heng*. On the basis of their pole-fixing characteristics, however, the extant discs that are not of modern fabrication would date, not from the Shang and Chou periods, but rather from approximately 300–400 C.E. Thus while the possible use of template discs for locating the celestial north pole in the absence of a convenient pole-star is not altogether to be dismissed, the extant discs that Price regarded as authentically ancient can nevertheless provide us with no information on the *hsüan-chi* and *yü-heng* of the *Shu ching*, which was committed to writing nearly a millennium earlier than their putative date.

Thus we do not know what the terms *hsüan-chi* and *yü-heng* might have signified to a Chinese of the mid-Chou period. One possibility, hinted at by some Chinese commentators (see Legge, *Chinese Classics*, III.1, 'Shoo King', p. 33), is that they were mythic names for celestial, rather than terrestrial, objects. Cullen, in

done any better? The instruments thus ordered were not only one or two, but quite a number; so that the results could be compared together. Such a wealth of equipment has never previously been recorded. All these His Majesty is intimately acquainted with, and even Kuo Shou-ching of the Yüan could have offered nothing better. After the Shou-shih calendar had been corrected, observational instruments (*kuan-t'ien chih ch'i* 觀天之器) were made, to follow the seasons of the heavens above and to be of service to the works of the people below.

His Majesty's sense of responsibility in the exploitation of the works of nature is of the highest, as also is his benevolence in the high valuation of agriculture. We Easterners have not seen anything as fine as these instruments before. Like the high tower of the Observatory itself, they will be passed down for time without end.

As these accounts make clear, the re-equipping of the Royal Observatory was undertaken at the personal initiative of the king, who oversaw the planning and execution of the project. He devoted substantial resources to it, and mobilised a

'Some Further Points on the *Shih*', p. 39, proposes that *hsüan-chi* (which, in one of several homophonic writings of that term, can be interpreted as 'rotating device') might refer to the pole-star, while *yü-heng* (the 'jade cross-piece') might refer to the Northern Dipper. This raises the further possibility that by *c.* the fifth century B.C.E. the term *hsüan-chi yü-heng* could have been a name for an astronomical instrument like the *shih* 式 ('diviner's board' or 'cosmic board'), in which a round 'heaven plate', pivoted at the the polar axis, was rotated so as to cause the seven stars of the Northern Dipper inscribed thereon to point in succession to the twenty-eight Lunar Lodges, the names of which were inscribed on a square base-plate (the 'earth plate').

Clearly this complex problem requires further research. Our feeling, however, is that, on the basis of present evidence, *hsüan-chi yü-heng* probably referred to an astronomical instrument of some sort, rather than to ritual implements of a non-astronomical character, at the time when the *Shu ching*'s 'Canon of Yao' and 'Canon of Shun' were written down.

Shun is said to have taken up the *hsüan-chi* and *yü-heng* so as to set in order the *ch'i cheng* 七政. The latter term, which can be understood to mean 'seven regulators' or 'seven (concerns of) government', also admits of several interpretations. In contexts where an astronomical meaning is clearly called for, it can signify either the seven stars of the Northern Dipper, or, collectively, the sun, the moon, and the five visible planets. In fact these two senses are closely related; if the *hsüan-chi yü-heng* was an instrument like the *shih* 'cosmic board', the seven stars of the Northern Dipper (*cheng* = 'regulators') would have been used to track or calculate the location amongst the Lunar Lodges of the sun, moon, and five visible planets (*cheng* = 'concerns of government' – because, in ancient China, positional astronomy and judicial astrology were inextricably linked). The later invention of the armillary sphere rendered the *shih* 'cosmic board' obsolete and, for the same reason, diminished in importance the role of the Northern Dipper as a celestial 'pointer'. Accordingly, in later Chinese astronomical texts the sense of *ch'i cheng* as 'sun, moon, and five visible planets' tends to predominate.

If the armillary sphere (*hun hsiang* 渾像) superseded the *shih*, it would have been natural for the alternative name *hsüan-chi yü-heng* to be transferred from one instrument to the other; and it now appears that that is precisely what happened. In any case, whatever *hsüan-chi* and *yü-heng* might have meant *c.* the fifth century B.C.E., by the first century C.E. they (taken together) were firmly established as a standard term for the armillary sphere (Needham, Wang, and Price, HC, pp. 17–18, 61, 122). From the Sung period onwards, the term *hsüan-chi yü-heng* was often used for an armillary sphere rotated automatically; as we shall see in Chapter 3, the term (Korean *sŏn'gi okhyŏng*) appears frequently in this sense in Korean accounts of the making of astronomical instruments in the middle Yi period.

large number of his officials to attend to it. The priority given the project is equated by the historian to that devoted to agriculture, traditionally the foundation of state policy in all of East Asia. The national commitment to the promotion of science and technology represented by this project probably exceeds anything that could be found elsewhere in the world in the early fifteenth century.

King Sejong and his court astronomers and other officials saw themselves as being heirs to a long and illustrious Chinese tradition of interest and innovation in the making of astronomical instruments. They placed their activities in a line of descent from the observations and instrument-making attributed to the (legendary) emperors Yao and Shun in high antiquity, through the more concrete achievements of the Han and T'ang and the comparatively recent triumphs of Kuo Shou-ching of the Yüan. Kuo-Shou-ching's Shou-shih calendar and his Simplified Instrument were completely familiar to them. Yet such was the confidence of Sejong's court scientists in their own successes that they considered themselves not only to reflect, but even to excel, the best that the Chinese tradition had to offer. The 'Men of the East', drawing on and surpassing their Chinese mentors, claimed to have assembled the best collection of astronomical instruments that the world had ever seen. Such a claim, bolstered by detailed descriptions of the instruments themselves, deserves our careful attention.

King Sejong's interest in timekeeping and astronomical instruments is mentioned as early as in an account of the making of a water-clock during the reign of his father, King T'aejong.[12] In 1424, Sejong, as king, ordered the casting of (the metal parts for) yet another clock, a Chinese-style Night-Watch Clepsydra (*kyŏngjŏmnu*; *keng-tien lou* 更點漏).[13] The king's interest in astronomy had apparently intensified by 1432, when he ordered the court debate on calendrical matters mentioned above, and the casting of a Simplified Instrument. Thereafter, the national policy of re-equipping the Royal Observatory was initiated; the work was substantially completed by 1439, although a few more instruments were made later. The instruments that are described in the official records as having been made as a part of this project are listed in Table 2.1.

In the remainder of this chapter, we present and discuss our translations of the

[12] Rufus, 'Astronomy in Korea', p. 29.

[13] *Ibid.* Rufus calls this instrument the *Kyung-chum-kui* (i.e. *kyŏngjŏm ki*; *keng-tien ch'i* 更點器). See also Jeon, STK, pp. 52, 56.

descriptions of these instruments as found in the *Sejong sillok* and elsewhere, following the order in which the instruments are listed in Table 2.1.

1. The Striking Clepsydra (*chagyŏngnu*; *tzu-chi lou* 自擊漏) (Figs. 2.1–5)

The fondness for clepsydras that King Sejong had shown in his youth culminated in 1432, when he commissioned the building of an elaborate instrument to announce the hours, intervals (*k'o*), and night-watches entirely automatically. The novelty and importance of this instrument (which was completed in 1434) are reflected, we believe, in the space devoted to a description of it in the Veritable Records of King Sejong. We shall begin by quoting that description in full, deferring our analysis until after the Yi historians have had their say.

Sejong sillok, 65: 1a ff; 16th year, 7th month (1434):

(On the first day of the eighth month 1434), the new clepsydra was brought into use. The king had decided that the old clepsydra[14] was not accurate enough, and had ordered the casting of (metal parts for) a new one, with four water-delivering dragon(-mouthed) vessels of different sizes. There were two inflow dragon(-ornamented) vessels, (one) for the double-hours and (one) for the night-watches, each 11.2 feet[15] long and 1.8 feet in diameter, having two indicator-rods 10.2 feet long.

The surfaces (of the rods) were divided into the twelve double-hours, each with eight intervals (*k'o*) and the extra fractions (*fen*) at the beginnings (*ch'u*) and mid-points (*cheng*) (of the double-hours) to make up (a total equivalent to) 100 intervals, each interval comprising twelve fractions.[16]

The night rods formerly (numbered) twenty-one, and were troublesome for the attendants to use in the night-watches. In conformity with the Shou-shih calendar's day/night apportionment increases-and-decreases, two (fortnightly) seasons were (now) served by one rod, (there being thus) altogether twelve rods [in the case of a non-striking

[14] Presumably King T'aejo's clepsydra of 1398 (see Ch. 1, p. 14 above), or else King Sejong's 'night-watch' clepsydra of 1424 (see p. 22 above, and also Jeon, STK, p. 56).

[15] The Chinese 'foot' (*ch'ih*; Korean *chŏk* 尺) was divided into ten 'inches' (*ts'un*; *ch'on* 寸), each further divided into tenths (*fen*; *p'un* 分). For the 'foot' used as a standard measure for astronomical instrument-making, see no. 17, p. 90 below.

Here and elsewhere, when dimensions of the Korean instruments are given in both *ch'ih* and *ts'un*, we translate the figures in decimal form; thus '2 *ch'ih* 5 *ts'un*' is translated as '2.5 feet'. Where dimensions are given in *ts'un* alone, we retain the translation 'inch', asking the reader to keep in mind that a Chinese 'inch' is a tenth, not a twelfth, of a Chinese 'foot'.

[16] For the complex systems whereby 100 intervals (*k'o*) were fitted into the twelve double-hours for time-annunciation purposes see Ch. 1, pp. 8–9 above, pp. 54–5 below, and also Jeon, STK, p. 87.

Table 2.1. *Equipment made for King Sejong's Royal Observatory, 1432–42*

	Name	Translation	Number of instruments made	Corresponding instrument of Kuo Shou-ching[a]	*Sejong sillok*	*Chŭngbo munhŏn pigo*
1.	*Chagyŏngnu* (*tzu-chi lou*) 自擊漏, in the Porugak (*Pao-lou ko*) 報漏閣	Striking Clepsydra, in the Annunciating Clepsydra Pavilion	1	—	65: 1a–3b	3: 1b
2.	*Ilsŏng chŏngsi ŭi* (*jih-hsing ting-shih i*) 日星定時儀	Sun-and-Stars Time-Determining Instrument	4	11, 12, 14	77: 7a–9a	—
3.	*Paekkak-hwan* (*pai-k'o huan*) 百刻環	Hundred-Interval ring	4	2[b]	77: 7a–9a	—
4.[c]	*So ilsŏng chŏngsi ŭi* (*hsiao jih-hsing ting-shih i*) 小日星定時儀	Small Sun-and-Stars Time-Determining Instrument	2 (several?)	11, 12, 14	77: 9a	—
5.	*Kanŭi tae* (*chien-i t'ai*) 簡儀臺	Simplified-Instrument Platform	1	—	77: 9b	—
6.	*Kanŭi* (*chien-i*) 簡儀	Simplified Instrument	1	2	77: 9b	—
7.	*So kanŭi* (*hsiao chien-i*) 小簡儀	Small Simplified Instrument	(several)	2	77: 9a, 10a	—
8.	*Kyup'yo* (*kuei-piao*) 圭表 and *yŏngbu* (*ying-fu*) 影符	Measuring-Scale and Gnomon, with a Shadow-Definer	1	5, 8	77: 9b	—
9.	*Honŭi* (*hun-i*) 渾儀	Armillary sphere	1[d]	1	77: 9b	—

	(Ch'in-ching ko) 欽敬閣	veneration				
12.	*Angbu ilgu* (*yang-fu jih-kuei*) 仰釜日晷	Scaphe sundial	2[e]	—	77: 10a	—
13.	*Hyŏnju ilgu* (*hsüan-chu jih-kuei*) 縣珠日晷	Plummet Sundial	(several)	15	77: 10a	—
14.	*Haengnu* (*hsing-lou*) 行漏	Travel Clepsydra	(several)	—	77: 10ab	—
15.	*Ch'ŏnpyŏng ilgu* (*t'ien-p'ing jih-kuei*) 天平日晷	Horizontal Sundial	(several)	—	77: 10b	—
16.	*Chŏngnam ilgu* (*ting-nan jih-kuei*) 定南日晷	South-Fixing Sundial	15 (10 of bronze)	—	77: 10b	—
17.	*Chuch'ŏk* (*Chou-ch'ih*) 周尺	Chou Foot-Rule	?	—	77: 11ab	—
18.	*Ch'ŭgu ki* (*ts'e-yü ch'i*) 測雨器	Rain-gauge	(several)	—	93: 22ab 96: 7ab	2: 32a

[a] Following the numbering of the list in Needham, *Science and Civilisation in China*, III: 369–70.

[b] This Hundred-Interval ring corresponds to the equatorial ring of the Simplified Instrument; see Fig. 2.14 and detail shown in Fig. 2.15.

[c] In Kim Ton's Memoir on the Simplified Instrument Platform, this and the next three instruments are given in the following order: 7, 4, 5, 6. The sequence has been changed here for the sake of a more orderly presentation.

[d] *Sejong sillok*, 60: 38b, records that additional armillary spheres of bronze (rather than of lacquered wood, as here) were made; unlike the present instrument, those were not mechanically rotated.

[e] The two scaphe sundials mentioned in our text were set up for public use; we assume that additional ones were made for palace and observatory use as well.

clepsydra].[17] When tested they agreed with the Simplified Instrument [cf. no. 6, below, pp. 64–6] without the slightest discrepancy.[18]

The king was also worried that the officials in charge of time announcements could not avoid mistakes,[19] so he ordered Chang Yŏngsil 莊英實 of the Palace Guard[20] to construct wooden jacks which would announce the time automatically without human agency.

The construction (was as follows):

First he erected a pavilion with three pillars (*ying* 楹). Between the eastern pair two storeys were built. On the upper storey stood three immortals as announcers, sounding double-hours by a bell, night-watches by a drum, and divisions of night-watches by a gong. On the storey below the middle was placed a horizontal wheel having twelve immortals round its circumference. Each immortal was carried on an iron rod upon which it could move up and down, and had a placard to announce one of the double-hours in turn.

The design of the mechanism for these movements (is as follows):

Between the centre pillars there is an upper platform which carries the outflow vessels, and lower down there are the inflow vessels. Above each of the (latter) vessels is erected a plain rectangular hollow wood(en casing) 11.4 feet long, 6 inches broad and 4 inches deep, (made of boards) 0.8 inch thick. Within these there are guides (*ko* 隔) projecting inwards rather more than 1 inch from the sides.

(In the casing on the) left is set a bronze rack as long as the indicator-rod but 2 inches wide, having twelve holes drilled in its surface to receive small bronze balls as big as

[17] With ordinary inflow-float clepsydras, multiple indicator-rods for night-watches (and correspondingly for the residual day double-hours) were necessary because the night-watches were equal divisions of the period from nominal dusk to nominal dawn, the length of which varied throughout the year. Twelve night-watch (and twelve day-double-hour) rods would serve for the twelve Fortnightly Periods (*ch'i*) of the first half of the year, and again in reverse order for those of the second half. An alternative system employed unique rods at the winter and summer solstices, so that the total requirement was thirteen night (and thirteen day) rods. Jeon (STK, p. 91) describes an eighteenth-century elaboration of this system in which each *ch'i* was divided into thirds so that the total requirement became thirty-seven night (and thirty-seven day) rods. The Striking Clepsydra needed only two rods (one for night and one for day), since the seasonal adjustments were effected by changing the night-watch ball-rack described in the succeeding passage of the text, but a full set of twelve night-rods might have been provided to allow visual time-readings (see p. 37 below).

[18] This important remark tends to confirm our suspicion that the Royal Observatory had a copy of Kuo Shou-ching's Simplified Instrument even before the re-equipping of the observatory began in the 1430s. See below, pp. 66–8.

[19] The old clepsydras were watched by Clepsydra Wardens, who had to announce the double-hours, and the night-watches and their divisions, with a complex system of signals on bells, drums, and gongs. See Jeon, STK, p. 56.

[20] Kim Pin's inscription says that Chang Yŏngsil was (formerly) a government slave in Tongnae 東萊 Prefecture. 'He was of an extraordinarily ingenious nature, always occupied with the work of artisans within the Palace.'

crossbow-bullets (*tan-wan* 彈丸). The ball-holes all have devices which can open and close, controlling the twelve double-hours.

(In the casing on the) right is set (another) bronze rack of the same length, but 2.5 inches wide, having twenty-five holes drilled in its surface to receive small balls like (those in the) left rack. Corresponding to the (usual) twelve (night-watch) indicator-rods (there are) altogether twelve (of these) racks, which in conformity with the Fortnightly Periods are used in rotation to control the night-watches and their divisions.

The float-rods of the water-receiving vessels[21] (function as follows):

The top of (each) rod raises (as it floats upward, a series of) horizontal iron (latches) like twigs on a stalk of bamboo (*chieh* 節), each 4.5 inches long. In front of the (inflow) vessels there are funnels (*hsien* 陷), to which are joined wide sloping boards the tops of which meet square holes (below the funnels) and the lower ends of which reach the eastern pillars.

Lower in the building there are four shelves (*ko* 隔) forming causeway-like channels (*yung-tao* 甬道). On (three of these) shelves rest iron balls as big as hens' eggs.[22] On the left, twelve (of these balls) control the (functions for the) twelve double-hours; in the centre, five control the (functions for the) night-watches and their initial divisions; while on the right twenty control the (functions for the remaining) divisions (of the night-watches). The ball-resting-places all have ball-releasers and -restrainers. Also there are horizontal devices which are shaped like spoons bent at one end to catch in a ring and rounded at the (other) end to receive a (bronze) ball. The waist of each spoon is mounted on an axle which allows it to go down and up. The rounded end of each spoon is aligned with an orifice in a bronze tube [i.e. one of a set of orifices along a bronze tube].

Of these bronze tubes there are two, set slanting above the shelves. The left-hand tube, 4.5 feet long and 1.5 inches in diameter, controls the double-hours; along its lower side twelve holes are bored. The right-hand tube, 8 feet long and the same diameter as the other, controls the night-watches and their divisions; along its lower side twenty-five holes are bored.

The holes all have devices which at the beginning leave every hole open. When one of the little balls from the bronze racks runs down it moves the device so as to close the hole

[21] In his brief descriptions of the Striking Clepsydra and its successors, Jeon (STK, pp. 58, 59) uses the expression 'water-receiving scoops' to translate the Chinese term *shou-shui hu* 受水壺. We presume his sources for this expression were Needham, Wang, and Price, HC, pp. 31 and 217, no. 44, and Needham, SCC, IV.2, Fig. 458, no. (3), where it was used for the buckets on the timekeeping waterwheels of the Sung astronomical clocktowers (see Chapter 1 p. 10, n. 24). In the Korean texts, the term *shou-shui hu* is that used for clepsydra inflow-float vessels, and we therefore translate it literally as 'water-receiving vessels'.

[22] Functions are specified for only three of the four shelves mentioned. We believe the fourth shelf held twelve still larger balls for operating the carousel jackwork described later.

automatically (after itself) and prepare a path for the next ball to run past (to the next hole), and so on one after the other.

Below the upper storey of the building between the eastern pillars (*ying*) there are suspended on the left two short tubes, one to receive the (iron) balls and the other to contain a mechanical spoon (set) so that its rounded end sticks out halfway across the bottom of the ball-receiving tube.

On the right (lower down) stand two round pillars (*chu* 柱) and two square pillars.[23] In the round pillars' hollow interiors are set devices like spoons half in and half out, five on the left-hand one and ten on the right-hand one. The square (hollow) pillars are slanting and traverse (the tops of) small tubes of which each one has four; (the small tubes) have lotus petals at one end and dragon mouths at the other. The lotus-petal (openings) are to receive the balls and the dragon mouths are to discharge them. The lotus petals and the dragon mouths face upwards and downwards respectively.

Above (on the right) there are also suspended two short tubes, one to receive the night-watch balls and the other to receive the night-watch-division balls.

The right-hand square pillar also has under each lotus petal two additional straight short tubes and one horizontal short tube. One end of this horizontal (short) tube is connected below a lotus petal of the left-hand square pillar. The five spoons of the left-hand round pillar, and also the five spoons of the right-hand round pillar, are so placed that their rounded ends each face between a dragon mouth and a lotus petal. The rounded ends of the five (other) spoons of the right-hand round pillar enter halfway into (the additional) straight (short) tube(s).

Thus as the water trickles down into the (left-hand) inflow vessel the indicator-rod floats upwards corresponding to the (passage of the) double-hours, and it opens (*po* 撥) the latches of the holes in the left bronze rack, so that the small (bronze) balls fall (one after the other) and find their way into the (4.5-foot-long) bronze tube. Falling through (its) holes, they operate the release mechanisms so that the large (iron) balls fall too and enter the (left-hand) short tube suspended under the upper storey; as they fall they operate the mechanical spoon, the other end of which transmits a motion from within the top of the (accompanying) tube to impel the elbow of the double-hour-announcing immortal (causing it to) sound the bell.

The watches' and divisions' (actions are) similar, but the night-watch (initial) balls run into the (central) suspended short tube to fall and operate the spoon mechanism, which from within the top of the left-hand round pillar impels the elbow of the night-watch-announcing immortal to sound the drum. (Then the balls) turn into the (night-watch)

[23] We have translated the word *chu* literally as 'pillar'. It is, however, clearly intended to be distinguished from the word *ying*, used for the structural pillars of the framework, and thus should be understood to mean 'chute', or the like (either vertical or sloping, as required).

divisions tube to operate the initial-division mechanism, which from within the top of the right-hand round pillar impels the elbow of the divisions-announcing immortal to sound the gong; then they stop.

At the entry turn to the straight small tubes below the lotus petals there is a device which initially closes the watch-balls' route (to the subsequent-division mechanisms) so that entry route is closed and the night-watch (mechanism) route is open. At (the initial divisions of) the other watches it is the same. When the fifth watch ends a shutter is withdrawn (and all the balls) come out.

Every ball for a night-watch division from the second onwards falls into the (right-hand) suspended short tube, enters the lotus petals, operates (its own) night-watch-division mechanism, and stops. Balls for later divisions turn past, operate (their own) night-watch division mechanisms, and stop.

The tubes where the balls stop have holes with shutters (*chiung* 扃) closing them, and when the ball for the fifth division falls it operates the lowest [i.e. the last] mechanism. Then an iron wire connecting the mechanisms pulls out all the shutters (so that this) ball and those of the previous three divisions all fall down at the same time.

(Each of) the large (iron) balls which control the double-hours falls into the (left) suspended short tube. Entering an additional round-pillar tube, it falls onto and depresses the northern end of a horizontal wooden lever, 6.6 feet long, 1.5 inches wide, and 1.7 inches thick.[24] At its centre this lever is pivoted on a round axle, upon a short pillar, so that it can rise and fall. At its south end stands a round wooden rod as (thick as) a finger and 2.2 feet long, which engages below the foot of (each in turn of the placard-bearing) double-hour-annunciating immortals. At the end of (each) foot there is a small wheel and axle (*lun chu* 輪軸). When the large ball hits the northern end, the southern end is raised and the foot of the immortal is impelled upwards.

Above the central storey, north of the lever's northern end, stands a small board which can open and shut. This board is connected by an iron wire with the spoon mechanism of the (left) suspended (short) tube controlling the double-hours. When the spoon mechanism is operated the board opens and lets out the front ball (from the fourth shelf).[25]

When the southern end of the lever drops, the double-hour-annunciating immortal returns to the face of the wheel, the next immortal (having) risen to replace it.

The wheel-rotation system (functions as follows):

[24] The apparent implication in this sentence that one of the hen-egg-sized balls itself goes on to operate the carousel jackwork is contradicted in the next paragraph, most of which we believe should logically be read before the sentence to which this note relates.

[25] We believe the 'front ball' to have been one of a set of twelve carried on the fourth shelf; see nn. 22 and 24 above.

Outside the wheel is set horizontally a small board about a foot long with a cutaway in the middle, for 4 or 5 inches, bridged horizontally by a bronze plate which has a pivot at one end to allow it to open and close.

The time-annunciating immortal's foot first enters about half an inch below the bronze plate. When it is raised it opens the bronze plate and goes above (the latter, which then) returns to the closed (position so that the immortal remains raised). When its double-hour is ended (the immortal) returns to the surface of the wheel, as (the horizontal wheel then turns so that) the iron wheel at the end of the foot runs along the bronze plate and drops off.[26] The next immortal works in the same way.

All the mechanisms are hidden so that they cannot be seen; the only things visible are the fully dressed immortals. This completes what one can say (about the Striking Clepsydra).

There follows the inscription of Kim Pin, parts of which we have already quoted above (pp. 17–18). Besides containing laudatory general remarks about the significance of the new clepsydra, emphasising its speed ('the mechanisms are tripped like lightning') and accuracy ('all who see it sigh with admiration at how extraordinarily it accords with (the motions of) Heaven'), Kim Pin's inscription provides a description of it that is very similar to the one just quoted, though with less technical detail. It adds that the Striking Clepsydra was installed in a new building called the Porugak (Pao-lou Ko 報漏閣; the 'Annunciating Clepsydra Pavilion'), and that bells, drums, and gongs were placed at the various palace gates, to be struck in response to the time-signals of the automatic clepsydra by Clepsydra Wardens from the Bureau of Astronomy. Thus relayed by human hands, the signals of the Striking Clepsydra became the standard annunciators of time for the entire palace and the city beyond.

The length and detail of the description of the Striking Clepsydra in the Veritable Records of King Sejong (incorporating verbatim Kim Pin's commemorative inscription) provide some evidence that it was regarded by the court chroniclers as being truly new and special. Many of the instruments mentioned in the *Sejong sillok*'s account of the re-equipping of the Royal Observatory are, as we shall see below, treated rather sketchily. It is difficult to know on what basis the chroniclers chose to reproduce in full, abridge, or omit the masses of source material that they had available for potential inclusion in the Veritable Records;

[26] Contrary to the implication of the (assumed) section-heading to this passage, the means of rotation are not described in the surviving text. We think there were auxiliary devices that rotated the carousel-wheel during the fall-back of the operating lever; see p. 38 below.

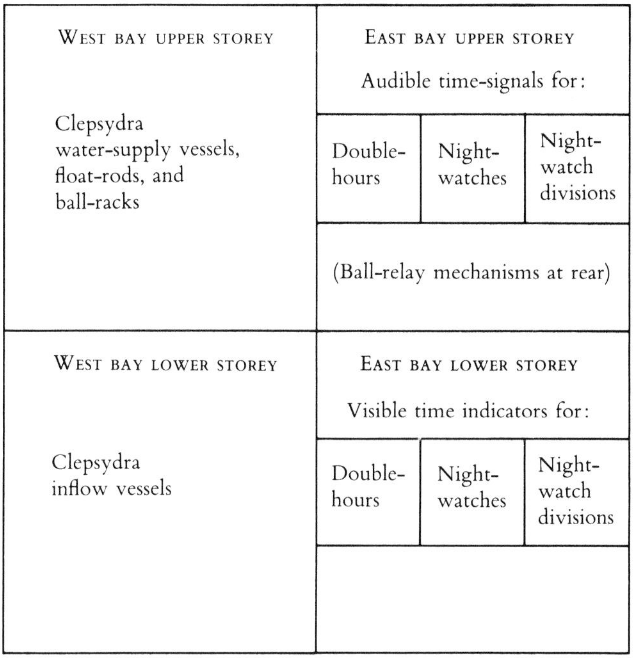

<table>
<tr><td rowspan="3">West bay upper storey

Clepsydra
water-supply vessels,
float-rods, and
ball-racks</td><td colspan="3">East bay upper storey
Audible time-signals for:</td></tr>
<tr><td>Double-
hours</td><td>Night-
watches</td><td>Night-
watch
divisions</td></tr>
<tr><td colspan="3">(Ball-relay mechanisms at rear)</td></tr>
<tr><td rowspan="3">West bay lower storey

Clepsydra
inflow vessels</td><td colspan="3">East bay lower storey
Visible time indicators for:</td></tr>
<tr><td>Double-
hours</td><td>Night-
watches</td><td>Night-
watch
divisions</td></tr>
<tr><td colspan="3"></td></tr>
</table>

Fig. 2.1. Striking Clepsydra: diagram of suggested south elevation.

we cannot even assume that their criteria were always well thought out. Nevertheless, in some cases it is not unreasonable to assume that the officials felt able to omit lengthy technical specifications for some of the instruments of the Royal Observatory, because their readers could be expected to be familiar with those instruments without the need for much descriptive detail. One has the impression in this case, however, that such an instrument as the Striking Clepsydra had never (as Kim Pin avers) been seen before, at least in Korea, and thus had to be minutely described. And while it is not quite true that the instrument is without parallels and antecedents, we may be grateful to the historians and memoir-writers for the attention that they devoted to it, for the description just quoted allows us to undertake a close analysis of this remarkable and long-vanished piece of time-keeping machinery.

Fig. 2.1 shows, in schematic fashion, how the various components of the Striking Clepsydra were arranged, with all of the works (except for the visible inflow vessels and display devices) hidden behind a wooden facade. As we shall describe in greater detail in Chapter 3, the clepsydra was repaired and a duplicate made in 1536; the original was destroyed in 1592 during the Hideyoshi invasions,

Fig. 2.2. Striking Clepsydra: surviving vessels of the 1536 copy of the original instrument, with restored tubes, float-rods, and stone platform, in the grounds of the King Sejong Memorial Hall, Seoul.

but the duplicate survived. Its jackwork, however, was rendered obsolete by a change in timekeeping methods in the middle of the seventeenth century, and accordingly was dismantled. The clepsydral apparatus itself has survived in part until the present time, although it was seriously damaged during the 1950–3 Korean War. Fig. 2.2 shows it as it appears today, in the grounds of the King Sejong Memorial Hall, Seoul.

As for the automatic striking mechanism, we may begin by repeating here the summary of its operation previously published in *Science and Civilisation in China*:[27]

> From two inflow clepsydra vessels there rose up two floating indicator-rods, one for the day and night double-hours, the other for the night-watches, dislodging as they did so 12 and 25 small bronze balls respectively from vertical bronze racks fixed above the clepsydra. The double-hour balls fell down one after another through appropriate conduits to dislodge 12 iron balls as big as hens' eggs; these then sped down a channel to make one immortal ring its bell, and at the same time, by means of unspecified additional mechanisms, rotate a horizontal wheel carrying 12 minor immortals with hour-annunciating placards so arranged that they appeared one after the other at a window. The system for the night-watch bronze balls was a little more complicated in that the tube in which they collected gave on to two separate channels with two rows of waiting iron balls, 5 to make the drum immortal operate and then to sound the gong for the beginnings of the night-watches, and 20 to mark the passage of the remaining fifths of night-watches by sounding the gong alone. An ingenious use of levers sufficed for nearly all the effects. In every case it was arranged that the detent orifices where the small balls tripped the large ones in succession should be closed automatically one after the other. At the end of the day and/or night the balls were all collected from the sumps where they had come to rest, and replaced at their starting points, no doubt when the water in the clepsydra vessels was changed.

Further study of the description of the Striking Clepsydra in *Sejong sillok* 65 allows us now to suggest in greater detail how its various mechanisms worked. Unfortunately, we have not been able to undertake the sort of large-scale modelling and experimentation that would be necessary for the production of drawings of its various sub-assemblies. Even if such practical studies were to be carried out in the future, as we hope they will be, we feel that working solutions for some parts of the mechanism will never be more than conjectural, owing to

[27] Needham, SCC IV.2: 517–18, modified.

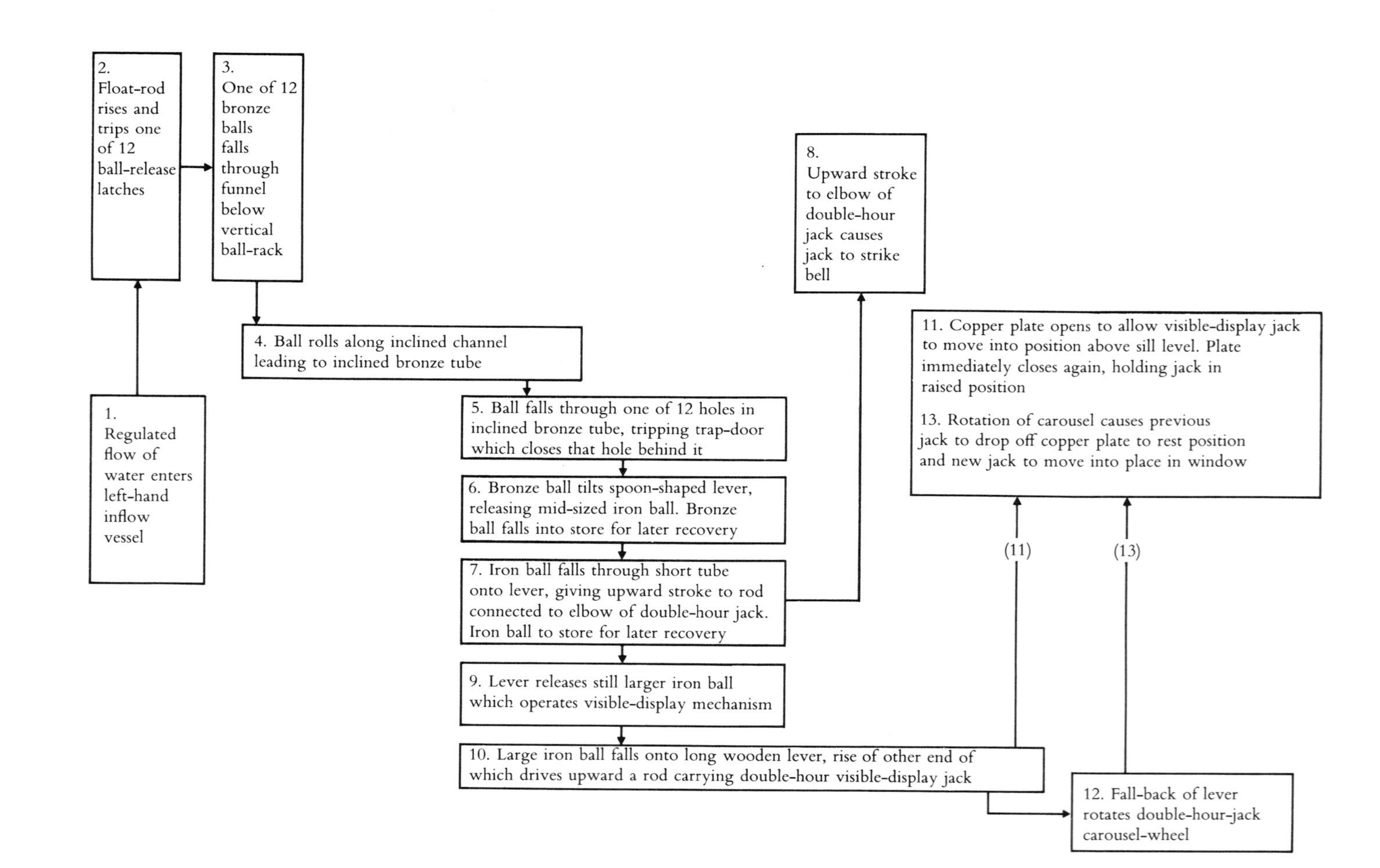

Fig. 2.3. Striking Clepsydra: double-hour annunciating system. Schematic diagram showing power-transmission path and sequence of operations.

a lack of information on some key points. Figs. 2.3 and 2.4 illustrate, in schematic fashion, the sequence of operations of the clepsydra's double-hour and night-watch mechanisms.

The information recorded in the passages translated above appears to have been obtained, perhaps with some omissions, from an original technical description written under what we assume to have been section-headings represented by the verbless phrases which in our translation we have completed by adding in parentheses '(was as follows):' etc.

We believe that, to house the components of which the dimensions are recorded or can be deduced, and to allow sufficient fall in the ball-runs between them, each of the two storeys of the Striking Clepsydra framework was perhaps some fifteen feet high. We suggest that each of its two south-facing bays also measured some fifteen feet in breadth and depth. The west bay housed the clepsydra vessels and ball-racks.

The upper storey of the east bay appears to have had in its south elevation three windows displaying the double-hour, night-watch, and night-watch-division audible time-signal jacks, each of which may have been life-sized. We think the lower storey of this bay also had three windows, of which that on the left was certainly for the double-hour visible time-indicating carousel jackwork; the other two were probably for night-watch and night-watch-division visible time-indicators.

The technical details can conveniently be considered sectionally as follows: clepsydra vessels; float-rod and ball-rack mechanisms; ball-relay mechanisms; visible time-indicator for double-hours; visible time-indicators for night-watches and night-watch divisions.

Clepsydra vessels

Four water-supply vessels are mentioned in the description of this 1434 Striking Clepsydra, in contrast to the three such vessels surviving as an apparently complete set from the 1536 instrument (Fig. 2.2, cf. Jeon, STK, Fig. 1.16 and p. 59). The fourth water-supply vessel of 1434 may have been another primary storage tank, affording a separate supply to each constant-level tank. Alternatively, if the medieval Chinese overflow method was used for water-level control – as in the Jade Clepsydra (pp. 77 and 79) – it may have been a receptacle for the overflow water from both constant-level tanks.

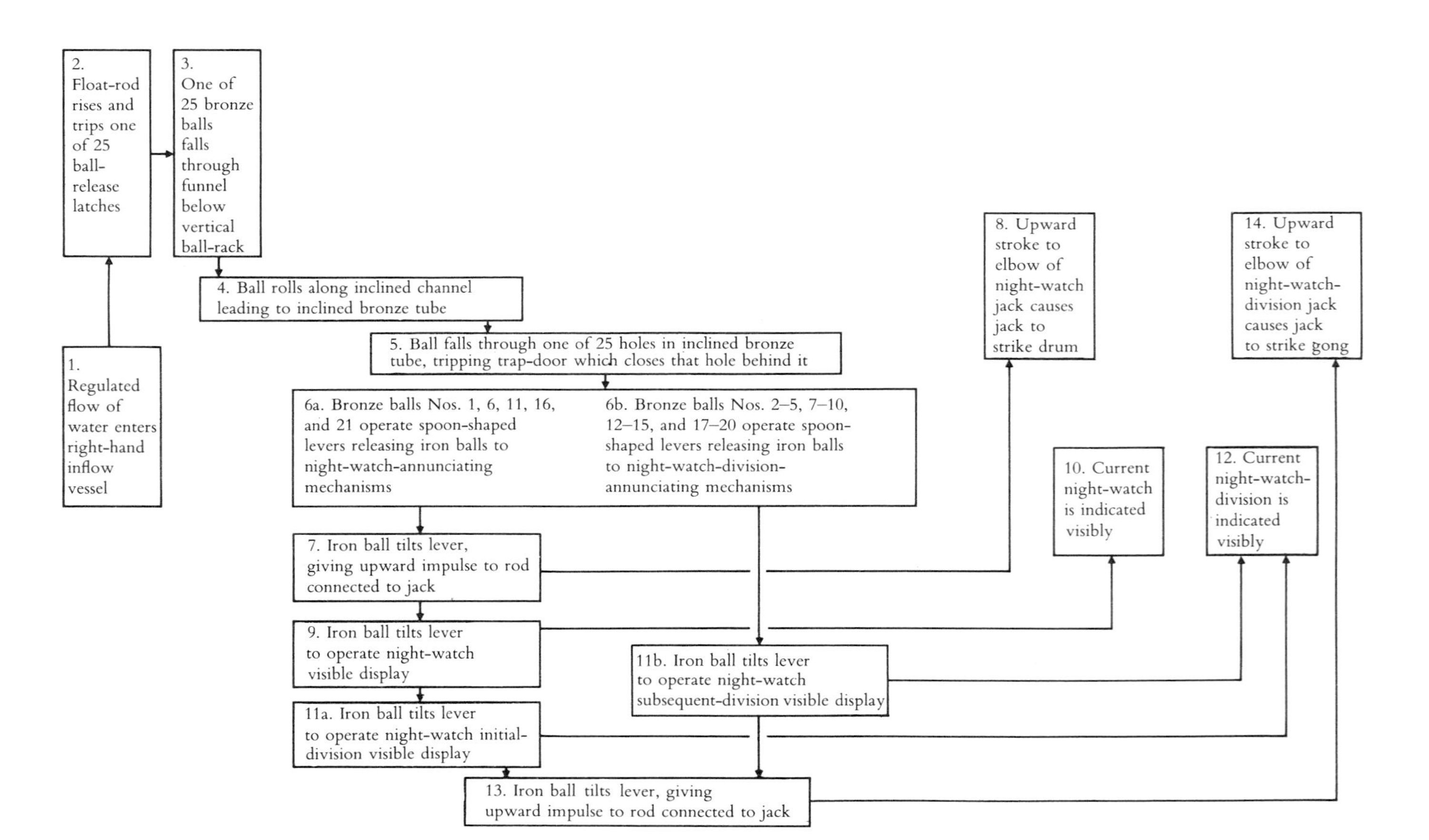

Fig. 2.4. Striking Clepsydra: night-watch annunciating system. Schematic diagram showing power-transmission path and sequence of operations.

Float-rod and ball-rack mechanisms

The float-rods rose through narrow wooden casings above the inflow-float vessels, and as they did so they raised horizontal latches which released small bronze balls from holes in vertical bronze racks.

For use with the float-rod of the night-watch inflow vessel, twelve interchangeable racks were provided. In each of these racks twenty-five ball-holes were spaced evenly along a length corresponding to the rise of the float-rod within the nominal duration of night for one Fortnightly Period in the spring half of the year, and for the corresponding period in the autumn half.

The text leaves us in some doubt as to whether the night-watch float-rods themselves also were changed fortnightly. There was no mechanical need for this, but it may have been done to allow the actual time at any moment of the night to be read by inspection of the float-rod.[28]

Where clepsydras were manually attended, it was customary to omit announcement of those double-hours that fell during the night-watches. In the Striking Clepsydra, however, the double-hour time-indicator was designed to be rotated through a complete cycle of twelve steps during each full day-and-night period. This suggests that double-hour announcements were continued during the night and that only one ball-rack, with twelve ball-holes spaced evenly along its effective operating length, was therefore needed for use with the double-hour inflow vessel.

Ball-relay mechanisms

Each of the twelve small bronze balls released from the double-hour ball-rack was caught by a funnel below the casing of the float-rod and ball-rack, and ran along an inclined channel leading to a 4.5-foot-long inclined bronze tube with twelve outlet-holes along its under-side. As each ball in turn fell through its own hole, it released a trapdoor which closed the hole after it to allow subsequent balls to pass to their proper holes further along the tube. It then operated a spoon-shaped lever which released one of twelve hen-egg-sized (say 1.5-inch-diameter) iron balls to the double-hour audible time-signal mechanism.

Each of the twenty-five small bronze balls from the night-watch ball-rack fell similarly through the correct one of twenty-five outlet holes from an eight-foot-

[28] See n. 17 above.

long inclined bronze tube. One ball in each five of this series (probably those balls occupying positions 1, 6, 11, 16, and 21) operated a spoon-shaped lever which released one of five hen-egg-sized iron balls along a route to operate first the night-watch and secondly the night-watch-division audible time-signal mechanism. Each of the other twenty balls in this series released one of twenty hen-egg-sized iron balls along a direct route to operate the night-watch-division audible time-signal mechanism. Besides operating the audible time-signal mechanism or mechanisms, each of the twenty-five hen-egg-sized balls also operated the appropriate visible time-indicating mechanism or mechanisms.

A supplementary ball-relay mechanism was operated by the double-hour time-signal mechanism, and served to release twelve even larger (perhaps 2.5- to 3.0-inch-diameter) iron balls used for driving the carousel jackwork of the double-hour visible time-indicator.

Audible time-signal mechanisms (Figs. 2.3–4)

The audible time signals were given by three wooden jackwork figures, perhaps life-sized, standing at the left-hand, central, and right-hand windows of the east bay upper storey. The jacks were designed to sound respectively a bell for double-hours, a drum for night-watches, and a gong for night-watch-divisions.

Each audible signal consisted of a single stroke, energised by an impulse transmitted to the jack's elbow by the impact of one of the hen-egg-sized balls upon the rounded end of a spoon-shaped lever at the lower end of a vertical tube.

The lever operating the double-hour bell also operated the supplementary ball-relay mechanism and thus initiated the operation of the double-hour visible time-indicator.

Visible time-indicator for double-hours (Fig. 2.3)

Twelve placard-bearing double-hour jacks, each perhaps about 0.5 feet tall, were supported by vertical rods on a horizontal carousel-wheel perhaps three feet in diameter. When each jack's double-hour arrived, the fall of the large iron ball from the supplementary ball-relay mechanism depressed one end of a 6.6-foot-long wooden lever, the other end of which raised the jack above the sill level of the viewing window, where a small wheel at its foot was then supported by a bronze retaining plate. By some unspecified means, such as perhaps a pawl-and-ratchet arrangement, the fall-back of the lever next caused the carousel-wheel to rotate and bring the jack to the centre of the window. The rotation of the carousel-wheel also caused the previous jack to run off the end of the retaining plate and so drop out of view.

Visible time-indicators for night-watches and their divisions (Fig. 2.4)

We believe these mechanisms provided visible indications of the commencement and duration of each of the five night-watches and those of each of the five divisions of every night-watch. In principle, this could have been done in analogue fashion by direct display of the relevant sets of one to five hen-egg-sized balls. The provision of three sets of five spoon-shaped operating levers suggests, however, that there was a successive display of tablets inscribed with the Chinese characters for the five night-watches (named for the first five of the Ten Stems (*kan* 干), viz. *chia* 甲, *i* 乙, *ping* 丙, *ting* 丁, *wu* 戊) and their five divisions.[29]

Five (probably the first, sixth, eleventh, sixteenth, and twenty-first) of the hen-egg-sized balls from the combined series of five-and-twenty were required, besides causing both the drum and the gong to be sounded, to operate both the appropriate night-watch and the initial-division visible displays; they were therefore routed through the relevant parts of the mechanism in succession. The four balls following each of these initial-division balls were required to operate only the gong and the appropriate subsequent-division displays.

Correct routing of each ball was probably effected by a system of runways with trapdoors or the like each of which was either self-closing like those in the ball-relay mechanisms, or closed by one of the ten spoon-shaped levers that worked the displays. When a trapdoor or display device needed resetting at a later stage in the cycle of operations, this would have been done either by another of the same levers or by one of the five additional levers.

The basic timekeeping mechanism of the Striking Clepsydra was a particularly

[29] Sources for the designations of the five night-watches include *Sui Shu* 隋書 19: 26b, *Sung shih* 76: 4ab, and *Yü Hai* 玉海 11: 10b. We have not traced any direct evidence for the visible designations used to identify the five divisions of each night-watch. At one time we thought they might have been the cardinal numbers 1–5, corresponding directly to the ordinal numbers used in the descriptive text. However, in the Sung calendars (see p. 55 and n. 47 below) the intervals (*k'o*) within double-hours are designated by the characters *ch'u*, *i*, *erh*, *san*, *ssu*, *wu*, *liu*, *ch'i*: i.e. 'beginning, 1, 2, 3, 4, 5, 6, 7'. In this series the first *k'o* is designated *ch'u*, while the second to eighth *k'o* are designated by the cardinal numbers 1 to 7. Though at first sight puzzling, this usage in fact corresponds exactly with that in modern Western digital clocks, where cardinal numbers 1, 2, 3, etc. designate hours which are ordinally the second, third, fourth, etc. after midnight. We think it is probable that the visible annunciator for night-watch divisions followed the same system, viz. *ch'u*, *i*, *erh*, *san*, *ssu*.

large and fine, but in principle quite ordinary, float-rod clepsydra of the sort that had long been familiar in China and Korea. The feature that so excited the enthusiasm of Kim Pin and his contemporaries was the use of that device to operate automatic jackwork time-annunciating mechanisms. While the particular arrangements of those mechanisms might well have been as innovative as Kim Pin claimed, we can nevertheless to some extent trace their ancestry.

The immediate inspiration for the Striking Clepsydra was probably the elaborate Chinese palace 'clock' built for the last emperor of the Yüan Dynasty, Shun-ti, in the 1350s and destroyed when the Ming armies sacked Peking in 1368. While extant descriptions of that device do not provide enough information to specify its power-transmission mechanisms, it does seem to have been in part a float-rod clepsydra.[30] Given the close relations between the Yüan and the Kingdom of Koryŏ,[31] it is virtually certain that Shun-ti's prized timekeeper would have been seen by Korean dignitaries visiting Peking. Complex timekeeping mechanisms were as much a symbol of the prestige of governments in medieval East Asia as great clocks were in medieval Europe; accounts of Shun-ti's 'clock' still circulating in Korea in the early Yi period might well have inspired so self-confident a monarch as King Sejong to commission an equivalent device.

Shun-ti's instrument in turn may be related to two mechanical clepsydras made for the first Yüan emperor, Kubilai Khan: the Precious Mountain Clepsydra (Pao-shan Lou 寶山漏), *c.* 1262, possibly constructed by Kuo Shou-ching,[32] and the Lantern Clepsydra (Teng Lou 燈漏) in the Grand Illumination Hall (Ta-ming Tien 大明殿), constructed in 1270 and definitely attributed to Kuo.[33] Of the former instrument we know nothing at all. The Lantern Clepsydra, described briefly in the *Yüan shih*, employed time-annunciating jackwork that externally, at least, must have resembled that of the Striking Clepsydra and also King Sejong's Jade Clepsydra (no. 11, pp. 76–80 below); in addition, it featured bejewelled

[30] Needham, Wang, and Price, HC, p. 140; Needham, SCC IV.2: 507–8. We no longer think, however, that this instrument was operated by a timekeeping waterwheel with a hydromechanical linkwork mechanism, as implied in the translation from *Yüan shih* 元史 43: 13b ff in HC and SCC IV.2 (*loc. cit.*). Shun-ti's palace 'clock' was certainly a clepsydra of some sort; we assume that, as in the Striking Clepsydra, it used float-rods and falling metal balls to transmit power to work its effects. Similarly, it is now clear, contrary to the suggestion in SCC IV.2: 519, that the Korean Striking Clepsydra and the Jade Clepsydra (no. 11, pp. 76–80 below) did not employ the waterwheel linkwork mechanism.

[31] Emperor Shun's second wife and empress was a Korean, the Lady Ki, daughter of a powerful Koryŏ aristocrat. She was a very active empress politically, and had close contacts with many Koreans (Dr Gari Ledyard, private comm.; see also Henthorn, *A History of Korea*, p. 128).

[32] *Yüan shih* 5: 2a; Needham, Wang, and Price, HC, p. 135, n. 6.

[33] *Yüan shih* 48: 7ab; Needham, Wang, and Price, HC, pp. 135–6.

lanterns as a part of its display apparatus. Its hidden mechanisms are not described, except that we are told that they were propelled by water; and that balls ('pearls') were used as a part of the time-annunciating system. Although we can say little about these presumed ancestors of the Striking Clepsydra and the Jade Clepsydra, their existence is very important, for they take us back to just the time when ideas of time-annunciating jackwork, remembered from the great Sung clocktowers, may have begun to be modified by new clepsydral techniques entering China from the Islamic world.

One feature of the Striking Clepsydra (and, by inference, of its Yüan predecessors) that particularly suggests inspiration from the Arabic tradition of water-clocks is the use of falling balls to transmit power. It is commonly thought that in Arabic water-clocks falling balls transmitted power only to the minimal extent that they were employed to fall onto bells, causing them to ring. This was true of the 'Archimedes' clepsydra,[34] and also of the famous chiming clepsydras at Antioch and Fez. But the third and fourth striking clepsydras described by al-Jazarī in his *Book of Knowledge of Ingenious Mechanical Devices* (1206 C.E.) provide a crucial exception to this rule.[35] In those, falling balls dropped onto bells in the ordinary way, but only after they had travelled through a complex pathway of levers and balances to operate jackwork figures as well (Fig. 2.5). The general similarity of the Lantern Clepsydra to the water-clocks described by al-Jazarī was noted by Needham, Wang, and Price.[36] Our detailed study of the ball-operated power-transmission mechanisms of the Striking Clepsydra's time-annunciating jackwork allows us now to suggest more strongly that its ancestry may be traced, through one or more Yüan intermediary instruments, back to the milieu of al-Jazarī himself.

There was a steady flow of mathematical and astronomical information between China and the Islamic world during the time of the *pax Mongolica* of the middle and late thirteenth century. Jamāl al-Dīn of the Marāghah Observatory, who explained Islamic astronomical instruments to Kuo Shou-ching, is the most

[34] Donald R. Hill, *On the Construction of Water-Clocks*, pp. 11–13, 19–20.

[35] Donald R. Hill, *Arabic Water-Clocks*, pp. 101, 111–19, esp. 116–18.

[36] Needham, Wang, and Price, HC, p. 135, n. 6.

André Sleeswyk has pointed out to us (private comm.) that the phrase 'balls as big as crossbow-bullets', common in the descriptions of Sino-Korean clepsydral devices, 'is reminiscent of Arabic usage of the word "bunḍuqa" for such metal balls in mechanisms, e.g. in al-Jazarī. . . . In general, the word indicated the bullet of the arbalest; it was derived from "buḍuq", hazel-nut.'

famous but by no means the only example of this.[37] It is quite possible that one of the Islamic mathematicians or astronomers who visited China at the time had a copy of al-Jazarī's treatise in his baggage; it is rather likely that one or more would have read the treatise and have been able to explain its contents to Kuo Shou-ching or one of his colleagues. Al-Jazarī's third and fourth striking clepsydras were developed directly from earlier Arabic ball-rung bell-devices, elaborated by ball-operated jackwork. (Conversely, possible Chinese influences on al-Jazarī's own work have been explored by Needham and others.)[38] A quite plausible route of transmission for al-Jazarī's ball-operated jackwork has been shown to have existed during the thirteenth century; such devices appear in full flower in King Sejong's Striking Clepsydra, and may be strongly suspected in the case of earlier Yüan instruments. These considerations in combination make it appropriate, we believe, to speak of a single Sino-Arabic clepsydral tradition during the Yüan period.

Arabic ball-mechanisms were, however, not copied directly in the construction of the Striking Clepsydra; we are dealing here with inspiration, not imitation. Arabic water-clocks such as those of al-Jazarī and the 'Archimedes' clepsydra were anaphoric, i.e. their floats were connected to weighted cords that rotated drums or axles. Their ball-release mechanisms therefore had a rotary element rather than being directly linear, as in the case of the Korean instrument. This transformation of the Arabic technique of dropping balls may reflect a general East Asian preference for float-rod clepsydras over anaphoric ones, and certainly the use of a *set* of periodically changeable ball-racks facilitated the timing of the East Asian variable night-watches. Whether this transformation took place in Seoul in the 1430s or whether, as now seems likely, it occurred in China in the 1260s or 1350s cannot, on the basis of information now available, be determined.

[37] Needham, SCC III: 372–4ff; on the transmission of mathematical information, see SCC III: 49–50.

[38] Needham, Wang, and Price, HC, pp. 187–91; Needham, SCC IV.2: 534–6ff.

André Sleeswyk has kindly reminded us (private comm.) that it would be appropriate at this point to mention Chang Heng's seismoscope of 132 C.E., which employed dropping, bell-ringing balls to announce the occurrence of earthquakes. The use of ball-dropping mechanisms for *event-marking* thus was known in China from early times, independently of any presumed later influence from the Islamic world. And time-marking is, as Sleeswyk points out, only a special case of event-marking.

It may well be the case, then, that ball-dropping mechanisms for event-marking were invented twice, independently – by Chang Heng and by the inventor of the 'Archimedes' clepsydra (although the possibility of some influence of the former on the latter deserves consideration). In that sense, the events described here might be regarded as the merging of two long-separated traditions. Nevertheless, it seems most likely that al-Jazarī provided the *immediate* inspiration for the bell-ringing clepsydras of China and Korea; those instruments descend only remotely from Chang Heng's seismoscope.

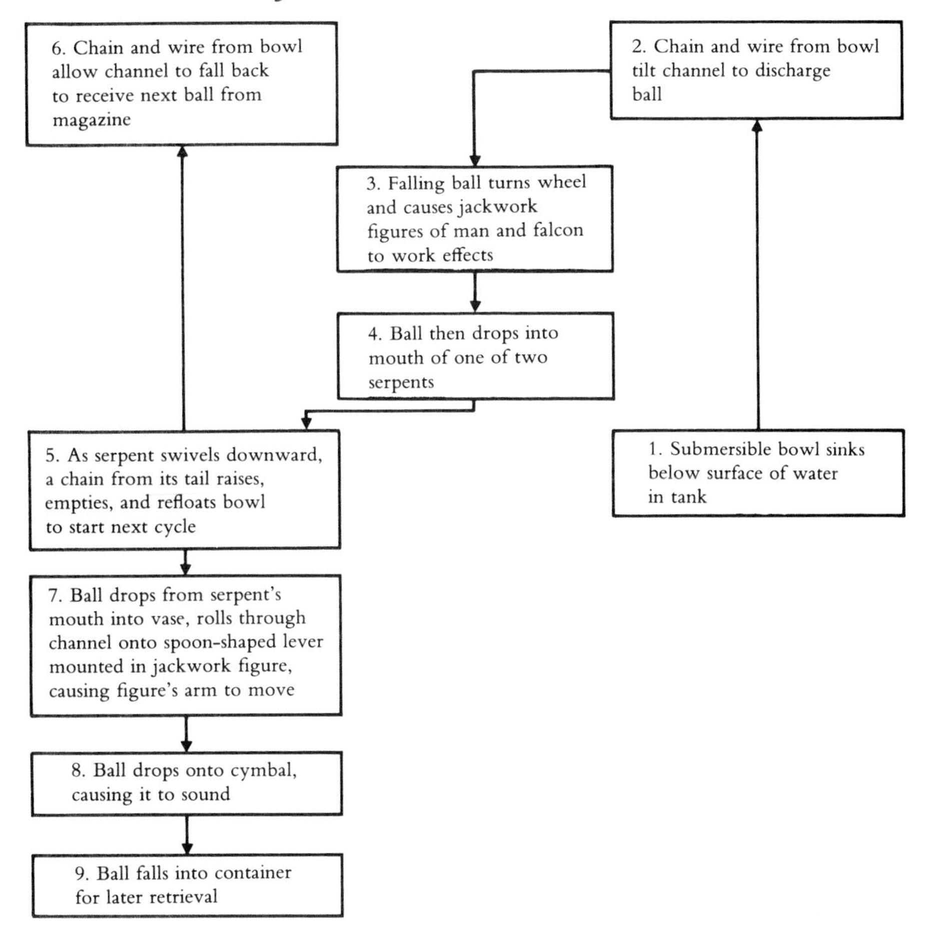

Fig. 2.5. Schematic diagram of ball-driven effects of al-Jazarī's fourth water-clock, for comparison with Striking Clepsydra. (Simplified; non-ball-driven effects omitted. For details see Donald R. Hill, *Arabic Water-Clocks*, pp. 112–19.)

Nevertheless, it is virtually certain that features of Arabic clepsydral technology could not have come to Korea by any route except through Yüan China. As we have noted, relations between China and the Kingdom of Koryŏ were very close during the reign of Yüan Shun-ti. Korean visitors to Peking who were shown Shun-ti's palace clepsydra during the ten years or so that it existed could, if they were careful observers with a sophisticated understanding of engineering, have figured out its methods of operation by themselves; but perhaps that assumes more technical interest and knowledge than one can reasonably expect of visiting dignitaries. It is also quite possible that Shun-ti's elaborate instrument had been the subject of an officially commissioned descriptive memoir or inscription similar to Kim Pin's inscription on the Striking Clepsydra. If a copy of this had been brought back to Korea by a visitor, it would have provided enough information to allow King Sejong's horological engineers to design an equivalent instrument nearly a century later. We can only guess that such a memoir or inscription might have existed, however; we know of no surviving copy, either in China or Korea, nor of any reference to such a document. Further, it is possible that Chinese horological engineers (or even technicians from further west in Asia) who had served at the Yüan court may have fled to Korea – still sympathetic to the fallen Mongol dynasty – after the Ming takeover in China.[39] There they could quite conceivably have preserved their specialised skills for the generation or two between the fall of the Koryŏ Kingdom and the commissioning of the Striking Clepsydra by King Sejong in 1432.

Whatever may have been the precise route of transmission of these techniques from the Islamic world through China to Korea, the fact that in the fifteenth century automatically sounding clepsydras employing falling balls were in use across the entire span of the Old World, from Fez to Seoul, testifies eloquently to the ability of ideas to travel far and quickly among the great civilisations of premodern times.

2. The Sun-and-Stars Time-Determining Instrument (*ilsŏng chŏngsi ŭi*, *jih-hsing ting-shih i* 日星定時儀), incorporating

3. The Hundred-Interval ring (*paekkak-hwan*; *pai-k'o huan* 百刻還) (Figs. 2.6–13)

The Sun-and-Stars Time-Determining Instrument was a compound instrument that functioned as both a sundial and a star-dial. The Hundred-Interval ring was an integral part of it; we also list this ring as a separate instrument, for reasons that will become clear below. The *ilsŏng chŏngsi ŭi* itself comprised an equatorial dial with three concentric bronze rings, of which the inner and outer were free to rotate relative to the fixed middle ring; an alidade; an axially mounted pair of columns bearing a double polar-sighting ring; and oblique threads running from the ends of the alidade to the outer polar-sighting ring, the whole being mounted atop a column fixed to a stand, which was equipped with a water-level (Fig. 2.6).

[39] This suggestion was made in Needham, SCC IV.2: 516.

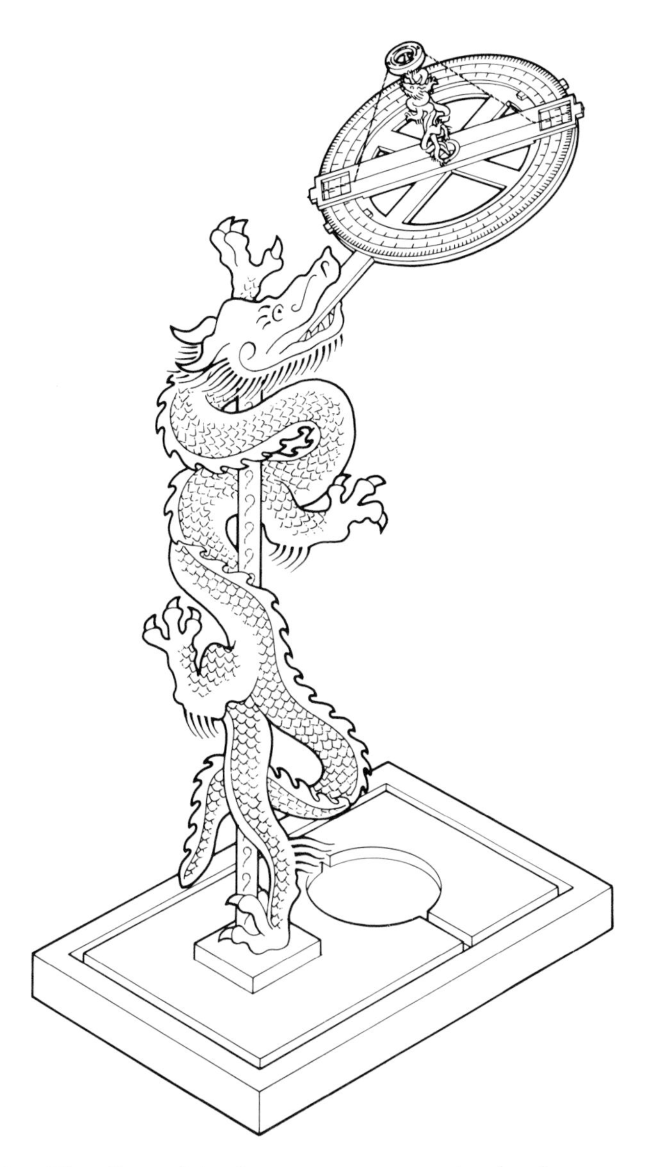

Fig. 2.6. Sun-and-Stars Time-Determining Instrument: reconstruction drawing.

The dimensions of the dial and associated parts of this instrument are recorded in sufficient detail to allow preparation of scale drawings, from which the perspective views were derived. The imaginative renderings of the large and small dragon-pillars are based on the tenth-century drawing of a dragon-pillar for the armillary sphere of Chang Ssu-hsün's astronomical clocktower of 979 C.E. (*Hsin i-hsiang fa-yao* 新儀象法要, 1: 19a; Combridge, 'Astronomical Clocktowers', Fig. 4).

The equatorial dial-plate (diameter 2 Chou feet; 490 mm) is supported in the plane of the celestial equator, and at a convenient height for use, by a dragon-column standing on a rectangular base (2 × 3.2 Chou feet; 490 × 785 mm) which has a pool and water-channels for levelling. The dial-plate carries three graduated scale-rings (see also Fig. 2.9) and a pivoted alidade (length 2.1 Chou feet; 515 mm) above which there are oblique sighting-threads strung from a polar ring held up by an axial pair of miniature dragon-pillars (each 1 Chou foot; 245 mm tall). The threads form a mobile gnomon for sundial readings by day, and for time-determinations by the stars at night, as explained in the captions to Figs. 2.7 and 2.9.

No example of the Sun-and-Stars Time-Determining Instrument survives, but there still exists an analogue to it, namely a pair of instruments which in effect dissect the compound instrument into its constituent sundial and star-dial. This pair of instruments is illustrated in Fig. 2.10, on which we will comment further below.

We begin our description of the *ilsŏng chŏngsi ŭi*, however, by quoting in full the lengthy and detailed description of it found in the Veritable Records of King Sejong:

Sejong sillok, 77: 7a–9a; 19th year, 4th month (1437):

At the beginning (of the year) the king had ordered the construction of instruments to measure the time by day and by night. They were called Sun-and-Stars Time-Determining Instruments. Four of these were made. One was decorated with clouds and dragons and placed in the royal palace. The other three had only a stand (*fu* 趺) to receive the handle (*ping* 柄) of the wheel, and straight columns to support the pole-fixing ring (*ting-chi huan* 定極還). One was given to the Royal Observatory for the purpose of observing the times (and seasons); the other two were given to the army headquarters of the Circuits of Hamgil 咸吉 and P'yŏngan 平安 for the purpose of improving vigilance in guard duties. The king commissioned Kim Ton to make an inscription . . .

We have already quoted the preface of this inscription above, pp. 18–19. Having made the obligatory general laudatory remarks, Kim Ton continues:

Now during the course of one solar revolution there are 100 intervals (*k'o*), half for the day and half for the night. For the daytime the reading of double-hours from the sundial is familiar. After dark, stars are used, as the *Chou li* 周禮 says, to divide the night, or again, as the *Yüan shih* 元史 says, to fix (the time), but they do not describe exactly the technique for doing it. For this purpose His Majesty decreed the construction of instruments to measure time by day and by night, which are called Sun-and-Stars Time-Determining Instruments.

The construction (is as follows): The instrument is of bronze. A wheel-shaped piece represents the equator. It has a protruding handle (*ping*); its diameter is 2 feet, its thickness 0.4 inches, and its width 3 inches. Spanning it there are cross-bars 1.5 inches wide, the thickness of which is the same as the wheel. In the middle of the cross there is an axle 0.55 inches long and 2 inches in diameter. (In the) north(-ern surface) of the axle a hole is counterbored so that only 0.01 inch of thickness remains. In the centre of this is drilled a small round hole no bigger than a mustard-seed. The axle fits into a hole in the alidade (*chieh-heng* 界衡; lit. 'boundary-beam'). The purpose of the (small) hole (in the axle) is to observe the stars.

Underneath there is set up a coiled dragon housing the handle (of the wheel). The protruding handle is 1.8 inches thick; it goes down into the dragon's mouth 1.1 feet and sticks out 3.6 inches. Under the dragon there is a platform 2 feet wide and 3.2 feet long, which has channels and a pool of water for levelling.

On the upper surface of the wheel there are three rings. One is called the celestial-circumference degrees-and-fractions ring (*chou-t'ien tu-fen huan* 周天度分還); the second is called the sundial Hundred-Interval ring (*jih-kuei pai-k'o huan* 日晷百刻還); the third is called the star-dial Hundred-Interval ring (*hsing-kuei pai-k'o kuan* 星晷百刻還). The first of these is on the outside and turns round; it has two ears (*erh* 耳). Its diameter is 2 feet, its thickness is 0.3 inch, and its width is 0.8 inch. The second of these is in the middle and does not rotate. Its diameter is 1.84 feet, and its thickness and width are the same as the previous one. The third one is innermost and turns round; it also has two ears. Its diameter is 1.68 feet, and its thickness and width are the same as the previous ones.

Above the three rings there is the alidade, 2.1 feet long, 3 inches broad, and 0.5 inches thick. Like the inner and outer rings, it has ears; the ears are used for rotating (the parts by hand). Both of its ends have rectangular slots 2.2 inches long and 1.8 inches wide so as not to obstruct (the visibility of) the graduations on the rings. At its centre, to the left and right (of the axle-hole) are two dragons each 1 foot long which hold up the 'pole-determining ring'. This is made of two rings, an outer one and an inner one, between which one can constantly see the great star of the Angular Arranger (*kou-ch'en* 勾陳). Within the inner ring one can see the Heavenly Pivot (*t'ien-shu* 天樞, i.e. the last star of the *pei-chi* 北極 constellation which begins with γ Ur. mi.). The purpose of this is to orient the polar axis and the equatorial plane. The diameter of the outer of these rings is 2.3 inches, and its width is 0.3 inch. The diameter of the inner ring is 1.45 inches, and its width is 0.04 inch [4 *li* 釐]. The thickness of both is the same, 0.2 inch. They are connected together by a little cross-piece. At both ends of the alidade, at each end of each slot, there is a little hole. The outer pole-determining ring also has small holes on both sides. Fine thread [or wire] is used to connect the six holes, linking the two ends of the alidade to the outer pole-determining ring. Thus above there is observation of the sun and stars, while below (there are means for) finding out the double-hours and intervals.

The celestial-circumference (degrees and fractions) ring is graduated with the [$365\frac{1}{4}$ Chinese celestial] degrees of the heavenly circumference, each degree being divided into four fractions (*fen*). The sundial ring is marked with the 100 intervals (*k'o*), each of which is divided into six fractions (*fen*). The star-dial ring is marked in the same manner as the preceding. But the midnight point [lit. the middle of the *tzu* 子 double-hour] each night oversteps (by a very small amount) the midnight point of the previous night, just as the rotation of the heavens produces an excess of one degree (annually) [$365\frac{1}{4}$ mean solar days = $366\frac{1}{4}$ sidereal days]. This is the only difference.

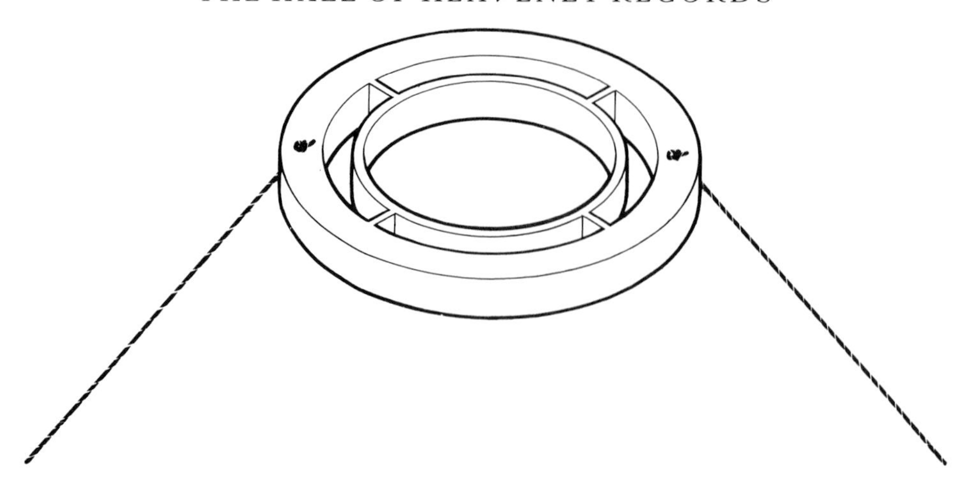

Fig. 2.7. Sun-and-Stars Time-Determining Instrument: detail of star-dial polar-sighting rings (reconstruction drawing).

When viewed from below through an axial sighting-hole in the pivot for the alidade (Fig. 2.6), the inner polar ring (internal diameter 1.37 Chou inch; 33.6 mm) circumscribed the apparent orbit of the T'ang Dynasty pole star T'ien-shu Hsing, and the outer polar ring (internal diameter 1.7 Chou inch; 41.7 mm) the apparent orbit of our present pole star α Ur. mi. as shown in Fig. 2.8.

For nocturnal time-determinations one of the sighting-threads was aligned with the star β Ur. mi. as shown in Fig. 2.8 and solar time was read on the innermost of the instrument's three scale-rings, as explained in the caption to Fig. 2.9.

The technique for using the celestial-circumference ring (is as follows): The clepsydra is used to determine the midnight point of the day preceding the winter solstice, and the alidade is used to determine the position of the North Pole (*pei-chi* 北極) constellation's second star [β Ur. mi.]. Then a record is marked on the side of the ring. This normally corresponds to the beginning of the first degree of celestial rotation. But after many years there will be a difference in the sidereal year (*t'ien-sui* 天歲). According to the Shou-shih calendar, after rather more than sixteen years there is a falling-back of one *fen* ($\frac{1}{4}^{d}$), and a whole degree in sixty-six years and a fraction. When this time comes it is necessary to re-determine the position. The second star of the North Pole constellation is near the north celestial pole (*pei-ch'en* 北辰), and as it is red and bright and easy to see, it is (the one that is) used (for time-determinations).

The function of the sundial Hundred-Interval ring is like that in the Simplified Instrument.

The function of the star-dial ring (is as follows): The beginning is at midnight preceding dawn on the first day of the winter solstice at the beginning of the first year. The beginning (of the first interval of the star-dial ring) is aligned with the beginning of the first degree of the celestial-circumference ring. (At the end of) one day (the star-dial ring is rotated so that it is aligned) with 1^{d}, at (the end of) two days with 2^{d}, at (the end

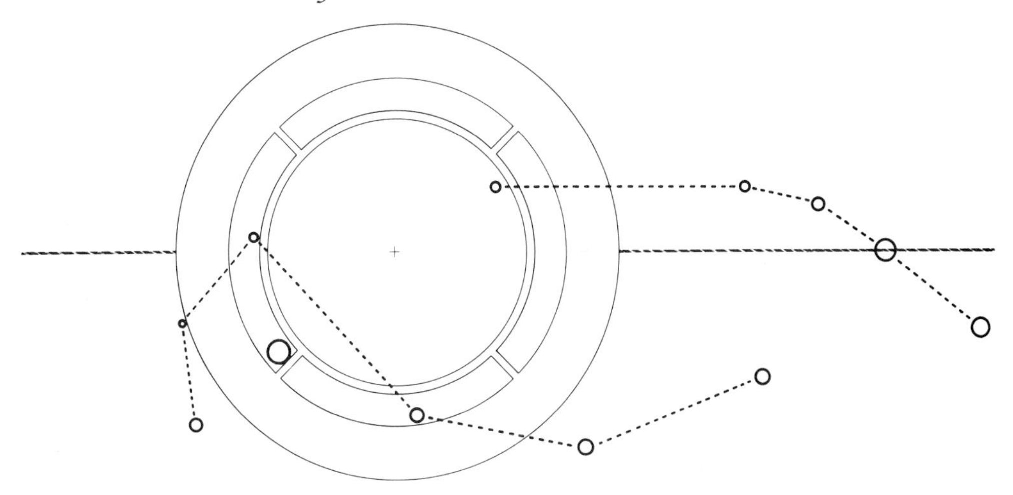

Fig. 2.8. Sun-and-Stars Time-Determining Instrument: relationship between star-dial polar-sighting rings and Chinese fifteenth-century circumpolar constellations.

The five stars of the Pei-chi ([Han to T'ang Dynasty] North Pole) constellation are shown above, and the six stars of the Kou-ch'en (Angular Arranger) constellation below, with the 'great star' α Ur. mi., our present Polaris, within its hook. (See Needham, SCC IV. 3, p. 566, n. a.) Small circles representing individual stars are drawn roughly according to visual magnitudes, but authorities disagree as to correct identification of the left-hand stars of Kou-ch'en.

The star-dial outer and inner polar-sighting rings are centred on the position of the north celestial pole of epoch 1410 C.E., and the right-hand sighting-thread is aligned on the second star, β Ur. mi., of Pei-chi.

of) three days with 3^d, and so on until 364 days are completed, when the alignment is with 364^d.

(At the beginning of) the next year, at the midnight point (before dawn on) the first day of the winter solstice, the setting will be 365^d. At one day (into that year, therefore) the setting will be zero and three *fen* (0.75^d); at two days the alignment will be with 1.75^d, and so on until 364 days are completed, when the setting will be 363.75^d.

The next year on the first day of the winter solstice the setting will be at 364.75^d. At one day (into that year) the setting will be zero and two *fen* (0.5^d). At two days the alignment will be with 1.5^d, and so on until 364 days are completed, when the alignment is with 363.5^d.

The next year on the first day of the winter solstice the setting will be at 364.5^d. At one day (into that year), the setting will be zero and one *fen* (0.25^d). At two days the alignment will be with 1.25^d, and so on until 365 days are completed, when the reading will be 364.25^d. This is called one completion (of the cycle of the residual fraction). When it reaches this position it starts all over again [i.e. in bringing the dial to the 365.25^d position for the start of the fifth year at the next midnight, a day is added automatically for the leap year].

Thus the mechanism of the activity and quiescence of humankind is related to the

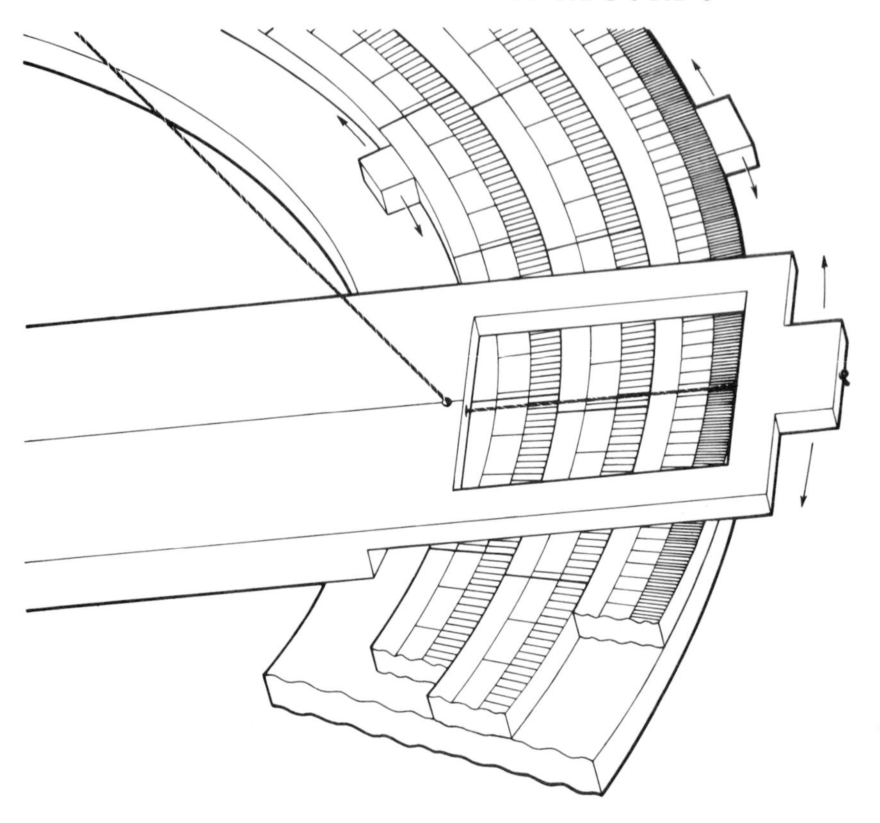

Fig. 2.9. Sun-and-Stars Time-Determining Instrument: detail of scale-rings, and of alidade with sighting-threads (reconstruction drawing).

The outer adjustable celestial-equator ring (external diameter 2 Chou feet; 490 mm) is graduated with $365\frac{1}{4}$ Chinese celestial degrees (*tu*), each of four fractions (*fen*). To make an allowance for the slow precessional movement of the equinoxes, with the celestial north pole, this ring was intended to be shifted 1^{d} anti-clockwise, as seen from above, every 66+ years.

The intermediate fixed celestial-equator ring (external diameter 1.84 Chou feet; 452 mm) is graduated for solar-time determinations, with 600 *fen* grouped in sixes for 96 of the 100 intervals (*k'o*) of the day-and-night, and in twenty-fives for the 24 half-double-hours, each of which contains $4\frac{1}{6}$ *k'o*.

The inner adjustable celestial-equator ring (external diameter 1.68 Chou feet; 412 mm) is graduated like the intermediate ring. To allow for the cumulative difference between solar time and sidereal time, and so permit solar time to be determined by star observations, this ring was intended to be shifted clockwise, as seen from above, from its datum position at midnight before the dawn of the first day of the winter solstice, by one celestial degree (*tu*) each midnight. Since there are $365\frac{1}{4}$ celestial degrees to a full circuit, the ring would have been moved through four full revolutions ($4 \times 365\frac{1}{4} = 1461^{d}$) only when $4 \times 365 + 1 = 1461$ days had passed, thus ensuring the automatic insertion of a leap-day every fourth year.

For daytime use the alidade was set so that the shadow of a sighting-thread fell along the centre-line of the alidade in the summer half-year, or midway between the pillars supporting the polar rings in the winter half-year; the time was then shown by the position of the thread on the intermediate scale-ring.

For nocturnal time-determinations the alidade was set so that when viewed from below, through the axial hole in the pivot for the alidade, a sighting-thread was aligned with the star β Ur. mi. as shown in Fig. 2.8; solar time was shown by the position of that thread on the inner scale-ring, provided the latter had been correctly adjusted for the date of the observation.

cyclical movements of the sun and the stars, such as are shown forth in the armillary sphere and the celestial globe. The sages of old made this the foundation of their method of government, like Yao attending to the calendar and the celestial sphere, and Shun examining the *hsüan-chi*. Now His Majesty's good intentions in causing these instruments to be constructed are of one mind with Yao and Shun; for a thousand years we Easterners have not experienced anything so grand. Such a wonderful achievement certainly deserves to be recorded for future generations, so I, Kim Ton, his servant, respectfully present the following inscription:

From the time of Yao and the calendar and globe, and Shun and the *hsüan-chi*, the generations have handed down the methods of making ingenious instruments, some called spheres and some called globes, not the same in appellation but always scrutinised for the guidance of the people's affairs. Now we are far from ancient times and although the methods and the records concerning them have been preserved, who can understand their meaning? But our sage ruler, responding to the ancestral techniques, has caused to be constructed the gnomons, the clepsydras, the spheres, and the globes of the Two Emperors (Yao and Shun). He restored the methods of old for regulating the double-hours and for using the Hundred Intervals to divide the day and the night. There was the sundial; but there was no one who did not wish (that it was possible to use such an instrument) also during the night. So He ordained that a new instrument should be made. What was its name? It was called the Sun-and-Stars Time-Determining Instrument.

How is it used? (The user) observes the stars with it, and correlates it with the sundial as follows: Its substance is of bronze; nothing can compare with it. A round wheel is first made, with north–south cross-struts. It is set in the plane of the celestial equator. A coiled dragon set onto the platform holds the protruding handle of the wheel in its mouth, and the platform has water-channels connected with a pool for levelling the main wheel. The three rings set on top of it are concentric. The outermost one is called the celestial-circumference degrees-and-fractions ring, and these (degrees and fractions) are graduated upon it. Within that there are two rings, the sundial ring and the star-dial ring, separately mounted. The graduations of the star-dial ring are larger than those of celestial degrees and fractions.[40] The outer and inner rings can both move; only the middle one is fixed fast.

An alidade lies horizontally on the face of (the apparatus), and has a pivot at the center. A hole is pierced through it, no bigger than a mustard-seed or the eye of a needle. The slotted ends of the alidade (point at) the degrees and intervals very precisely. Two dragons on each side of the pivot hold up the pole-determining rings. These consist of an

[40] The star-dial ring was divided into 100 intervals (*k'o*) each of 6 fractions (*fen*), i.e. into a total of 600 fractions, while the celestial-degree circle was divided into 1461 quarter-degrees.

outer and an inner (ring), between which stars are seen. What stars are these? They are the (great) star (of the constellation) Angular Arranger (*kou-ch'en*) and the Heavenly Pivot (*t'ien-shu*) star, used to locate the polar axis.

How are *mao* 卯 and *yu* 酉 [Heavenly Branches used as directional markers for east and west] fixed corresponding with their proper double-hours? (This is done by) using threads stretched from the (pole-determining) ring above and attached to the ends of the alidade below.

The method for measuring the (motion of the) sun involves two (threads); that for observing the stars needs only one. The red and bright star of the constellation Imperial Throne (*ti-tso* 帝座) is near the celestial pole, and by using a thread one can know the pole itself, and the intervals. The clepsydra being used to fix the midnight (moment), this is marked on the wheel; this is where the ring's circuit of the heavens begins. Every night it passes through the celestial degrees and fractions, from beginning to end.

The instrument is simple but refined. Its use is all-embracing and full of detail. Its mechanism is quite classical, like that devised by the philosophers of old. What is missing it supplies. Now His Majesty has begun its construction with a creative spirit like that of Hsi and Ho, having himself designed this precious thing. It is made of bronze, and when it reaches the moment of completeness, it starts all over again.

It has been made under the supervision of His Majesty himself. He gave the orders to the Transmitter of Directives Kim Ton, and to the Auxiliary Academician Kim Pin, who said, 'I do not dare to write an essay on this, but I hope that you higher officials will make an inscription according to the account which I have given herewith.' The account is given here in order that it may be preserved for posterity. His Majesty's wish to have a detailed statement about the system for the determination of time in a simple and easy manner is here shown as clearly as the palm of the hand. We, Kim Ton and others, cannot presume to change a single word of it, but only add an introduction and a conclusion. The inscription therefore has been (written) in this way.

The Sun-and-Stars Time-Determining Instrument is derived directly from the mobile equatorial ring and associated components of Kuo Shou-ching's Simplified Instrument;[41] this is apparent from the design and operation of the instrument, and is reinforced by Kim Ton's statement that 'the function of the sundial Hundred-Interval ring is like that in the Simplified Instrument'. The equatorial ring of the Simplified Instrument was moveable, like the star-dial ring of the

[41] See the schematic drawing in Needham, SCC III: 371, Fig. 166, letter 'j', and also our Figs. 2.14 and 2.15. See also Jeon, STK, p. 73.

Fig. 2.10. Paired sundial and star-dial, which taken together, are analogous to the Sun-and-Stars Time-Determining Instrument.

present instrument; and it was equipped with two alidades (which were also called *chieh-heng*, 'boundary-beams'), fitted with oblique wires.[42]

Moreover, as shown in Table 2.1, the *ilsŏng chŏngsi ŭi* apparently had a direct counterpart among Kuo Shou-ching's own instruments of the 1270s; that is, Kuo seems also to have used the equatorial ring of his Simplified Instrument as the basis for an independent device. The 'star-dial' (*hsing-kuei* 星晷), 'time-determining instrument' (*ting-shih i* 定時儀), and 'pole-observing instrument' (*hou-chi i* 候極儀), though listed separately in the *Yüan shih*, are probably all components of a single instrument. The first two terms might in fact be synonyms, while the third refers to pole-fixing rings, presumably mounted above an alidade. Thus Kuo Shou-ching seems to have made an instrument substantially identical to the star-dial shown on the right in Fig. 2.10. The innovation represented by the *ilsŏng*

[42] Needham, SCC III: 371, Fig. 166, letters 'k, k′'; see also A. Wylie, 'The Mongol Astronomical Instruments in Peking', in his *Chinese Researches*, part III, p. 10.

chŏngsi ŭi was that it combined that sort of star-dial with an equatorial sundial, as described above. Once again we find support for the claim of King Sejong's astronomers that they did not merely duplicate the old instruments, but also extended and improved upon them.

The equatorially mounted Hundred-Interval ring, a miniature of Kuo Shou-ching's equatorial ring, is the essential reference-point of the Sun-and-Stars Time-Determining Instrument. The scale of this ring is of considerable interest, for it emphasises the derivation of this instrument from those of Kuo Shou-ching. As can be seen from Fig. 2.9, the scale-ring is divided into twelve double-hours, each half of which comprises 25 *fen*, for a total of 600 *fen*. Of these, 576 are grouped in sixes to form 96 complete intervals (*k'o*), while the remaining 24 fractions (*fen*), representing the missing 4 intervals (*k'o*), are distributed so as to appear individually at the ends of each of the twenty-four half-double-hours. This scaling of the *paekkak-hwan* explains the reference in the description of the Striking Clepsydra (p. 23 above) to extra fractions (*fen*) at the beginnings and mid-points (of the double-hours).

The 600 *fen* of this ring correspond exactly to the 600 teeth of the timekeeping gear-wheel of Su Sung's 1088 astronomical clock,[43] because the same arithmetic applies to both. Their distribution amongst 96 *k'o* is probably the key to Su Sung's description of the equatorial scale of his armillary sphere as having 'the twelve (double-)hours with the division into "quarter-hours" [intervals, *k'o*] and "minutes" [fractions, *fen*] ... from the beginnings and middles of the hours so as to make the (full) number of 100 [intervals]'.[44]

Further, Su Sung's clock had 96, not 100, jacks for announcing the *k'o*. Combridge assumed earlier that they announced *k'o* reckoned in unbroken succession from midnight, omitting just the four *k'o* announcements which would then have coincided with those for the mid-points of the six a.m., midday, six p.m., and midnight double-hours;[45] but later he took the view that they announced *k'o* which were reckoned anew from the beginning 'and perhaps from the middle' of each double-hour.[46] In *Sung shih* 宋史 Chapters 70 and 76 there are

[43] Needham, Wang, and Price, HC, p. 36; Combridge, 'The Astronomical Clocktowers of Chang Ssu-hsun and his Successors', p. 299.

[44] H. Maspero, 'Les instruments astronomiques des chinois au temps des Han', *Mélanges chinois et bouddhiques*, 1938–9, *6*: 183–370, p. 310; authors' tr.

[45] Combridge, 'The Celestial Balance', p. 83.

[46] Combridge, 'Clocktower Millenary Reflections', p. 605.

tables dating from 1010 C.E. and 1049 C.E. which give sunrise and sunset times in terms of whole *k'o* reckoned from the middle of a double-hour towards and past the middle of the next double-hour, plus fractions of a *k'o*.[47] It seems possible that the placards carried by the *k'o* jacks followed this system of reckoning *k'o* exclusively from the mid-points of double-hours, and not anew from their beginnings, but unfortunately the text describing Su Sung's clock is silent on this matter.

The power-source of Su Sung's clock was set so as to make six 'ticks' per *fen*, a total of 3600 'ticks' for the 600-tooth timekeeping wheel to make one complete revolution in a day.[48] This figure is in turn probably the source of the division of the equatorial ring of Kuo Shou-ching's Simplified Instrument into 3600 *fen*. While the Hundred-Interval ring of the Sun-and-Stars Time-Determining Instrument was derived from that equatorial ring, its much smaller size would have forced a scaling-down of the 3600 *fen* to the 600 originally used for timekeeping purposes. Two-minute divisions probably would have approached the limit of accuracy with which the movement of a shadow could be read on such a relatively small ring.

The equatorial ring of the Simplified Instrument had two alidades. In the instrument described here, the single alidade carried two upright dragon pillars which supported a concentric pair of rings (Figs. 2.7–8) used to sight-in a group of circumpolar stars. From the outer ring threads ran to each end of the alidade, fixed in small holes on each end of the slot through which the scales of the rings were read. The alidades of the Simplified Instrument also carried oblique wires. This apparently was the usual practice; later Chinese and Korean equatorial sundials had oblique threads (see Fig. 2.13; compare Figs. 2.18–19), and another surviving Hundred-Interval ring from the Yi period had a thread attached centrally on each end of the alidade. Fig. 2.11 is a detail of this ring; it is similar to the sundial ring on the left in Fig. 2.10, but it lacks a protruding handle, while its alidade survives.

In the Sun-and-Stars Time-Determining Instrument, the oblique threads per-

[47] *Sung shih* 70: 25a–26a, 76: 4b–7a. The fractions in the former are $\frac{1}{150}$ ths of a *k'o*, and in the latter $\frac{1}{60}$ ths. The *k'o* are in both texts reckoned from the *middles* of double-hours, though in Chapter 70 this is not made explicit; and they run in an unbroken sequence which clearly extends to $8\frac{1}{3}$ *k'o* in a double-hour, though actual entries of 8 *k'o* or above do not appear in the tables because of rapid changes of sunrise and sunset times near the equinoxes. The *beginning* of the next double-hour, however, became due at $4\frac{1}{6}$ *k'o* from the *middle* of the one named, and the fact that the *k'o*-numbering (see n. 29 above) continued in unbroken sequence beyond this new beginning probably accounts for a remark that the two double-hours 'overlapped with each other almost by half' (*Sung shih* 76: 4a, tr. in Needham, Wang, and Price, HC, p. 93, n. 5).

[48] Combridge, 'The Celestial Balance'.

formed several functions. They could be set to mark the meridian, and used to observe the culmination of stars. They would also have been used in conjunction with the axial pin-hole and with the pole-fixing rings to sight-in on appropriate stars to aim the instrument properly. Their most important function, however, was to allow the determination of time by both day and night.

The fixed equatorial sundial ring was used with the alidade and the oblique threads to constitute a sundial of unusual accuracy. The oblique threads were in this case simply a mobile gnomon; one had only to move the alidade until the shadow of a thread lay centrally along it, and the alidade would point to the correct time on the Hundred-Interval ring. Of course, this shadow-casting method would only work during the six months of the year, from March to September, when the sun was above the equator and could shine on the face of the ring and its alidade. The mounting of the oblique threads was, however, an ingenious way of overcoming this basic problem of equatorially mounted sundials. (For the more common solution to this problem, see below, p. 88 and n. 93.) For even when the sun was below the equator, the centrally mounted pillars bearing the pole-fixing rings and the threads attached to them would remain in the sunlight, being sufficiently high above the shadow cast by the dial itself. Thus in winter the alidade would be turned until the shadow of a thread fell evenly through the slot between the central pillars, when the correct time could be read from the scale visible in the alidade's slotted end.

The accuracy of this sophisticated sundial was such that, as Nakayama reports,[49] a version of it (similarly called the *hyakkoku kan* 百刻還) was used in the seventeenth century by the great Japanese calendar reformer Shibukawa Harumi 澁川春海 as the standard timekeeping instrument for calendrical purposes, replacing the clepsydra that had been used previously to determine standard time.

The fixed sundial Hundred-Interval ring was used independently of the moveable celestial-circumference ring; that ring and the star-dial Hundred-Interval ring were used together. The celestial-circumference ring, graduated in $365\frac{1}{4}^{d}$, was moveable, but in fact was rarely moved: it was designed to be moved anticlockwise only 1^{d} every 66+ years, to make an allowance for the slow precessional movement of the equinoxes and celestial north pole. This instrument was obviously expected to remain in use for a very long time.

[49] Nakayama, *A History of Japanese Astronomy*, p. 123.

Fig. 2.11. Detail of yet another Yi period 'Hundred-Interval ring' sundial , similar to that in Fig. 2.10. left, showing graduation in double-hours and half-double-hours, 96 intervals (*k'o*) each of 6 fractions (*fen*), and 'remainder *fen*' of the other 4 intervals distributed as 1 *fen* per half-double-hour.

Once the celestial-circumference ring was correctly orientated, the star-dial ring could be used with it. The star-dial ring was moved 1^d clockwise per day relative to the celestial-circumference ring, to account for the cumulative difference between solar and sidereal time. The fact that the celestial-circumference ring was marked with $365\frac{1}{4}^d$ ensured that a leap-day was added automatically every four years. With the hour-scale of the star-dial thus kept truly oriented to the stars, the time could be told at night by using the axial sighting-hole and the oblique threads of the alidade to sight-in on β Ur. mi., allowing solar time to be read (through the slot in the alidade) on the star-dial Hundred-Interval scale.

The star-dial function of the Sun-and-Stars Time-Determining Instrument was of course exactly analogous to that of the sixteenth-century (1520 onwards) Western portable Nocturnal Dial (or simply 'Nocturnal'),[50] though with the refinements of a more accurate sighting method, and adjustability for precession. In the Western version the scales are on the south-polar face of the dial, and therefore run anti-clockwise for both solar-time readings and adjustments of date. The presence of such basically similar instruments at both extremities of the Eurasian continent raises questions of technique-transmission which we shall not pursue here, beyond mentioning the possibility of a shared descent from the Simplified Instrument or some forerunner thereof.

As may be seen by comparing our reconstruction drawing of the *ilsŏng chŏngsi ŭi* (Fig. 2.6) and the photograph of a paired equatorial sundial and star-dial that survived nearly intact until the 1950–3 Korean War (Fig. 2.10), the latter instruments represent a simplified dissection of the former one. The sundial is missing its alidade, central pillars, and oblique threads; but we know from another surviving *paekkak-hwan* sundial that they must have existed.

The lately surviving dials pictured in Fig. 2.10 were simple in form and devoid of ornamentation. The base of the sundial bears an inscription which reads, in part, 'Royally commissioned instrument. A time-measuring instrument, on which the beginnings and middles (of the double-hours) are inscribed in equalised degrees (*chün-tu* 均度). Its gnomon is straight and its shadow true; it conforms to the Way of the Ruler ...'

As the instruments appear in the photograph, the rings are not elevated on columns, as the *ilsŏng chŏngsi ŭi* was described as having been; instead their handles are simply dovetailed into slots in their water-level bases. But we doubt that such was the original arrangement. With the dials near ground-level it would have been difficult to read the sundial, and impossible to tell the time by the stars or even obtain a polar fix for the star-dial. Even had the bases themselves been placed on plinths of convenient height, the mounting of the rings would have remained awkward for use. Moreover we deduce from the photograph that, when it was taken, the rings as dovetailed into the slots were at about 25° elevation, instead of

[50] Needham, SCC III: 338 and Figs. 152–3; F. A. B. Ward, *Time Measurement, Part 1, 'Historical Review'* (1st edn, London, 1936), p. 22 and Pl. V; H. O. Hill and E. W. Paget-Tomlinson, *Instruments of Navigation* (London: National Maritime Museum, 1958), pp. 50–4 and Fig. 9.

the correct (for Seoul) 52.5°, and the polar axis accordingly was at about 65° instead of the correct 37.5°. We conclude therefore that the mounting arrangement shown was not the original one, and that the rings were probably originally elevated above the water-level bases on columns set into the surviving dovetail slots.

We also note that the equatorial circle of Kuo Shou-ching's Simplified Instrument, as elaborated in the Sino-Korean tradition of equatorial sun- and star-dials with alidades, central pillars, and oblique threads, had yet another apparent descendant – this time among the seventeenth-century instruments of the Chinese Jesuit fathers (Fig. 2.12). This was the Azimuthalem Horizontem (*ti-p'ing ching i* 地平經儀) of Ferdinand Verbiest, S. J. (Nan Huai-jen 南懷仁, 1623–88, the successor to J. A. Schall von Bell as the second European Director of the Board of Astronomy and Calendar under the Ch'ing Dynasty). The alignment of this instrument is towards the horizon and the zenith; its horizontal circle is graduated in 360 Western degrees. That is, its mounting and graduation are prima facie European, but it is nevertheless markedly Chinese (or Sino-Korean) in conception. Verbiest appears to have taken the idea for this instrument from the Simplified Instrument or one of its derivatives, rather than from Tycho Brahe as was the case with most of his instruments.[51] Tycho of course stressed equatorial-polar alignment in his instruments; Verbiest here apparently borrowed the design of a Chinese equatorially mounted instrument and changed its alignment according to older European practice. The use of oblique threads as seen in the Azimuthalem Horizontem thus may have entered European instrumentation in general as a direct borrowing from Chinese practice.

Jeon says that the use of thread-gnomons was introduced into East Asia from the Islamic world,[52] but that statement would seem to be true only in the sense that the oblique threads from the alidade of the Simplified Instrument may have been derived from Arabic instruments. Within East Asia, all thread-gnomons on subsequent instruments appear to have been derived directly or indirectly from those of the equatorial Hundred-Interval ring of the Simplified Instrument itself. We shall comment further on this below, in connection with the equatorial Plummet Sundial and the Horizontal Sundial (nos. 13 and 15; Figs. 2.18 and 2.19). Here it will be appropriate to describe in detail a late type of equatorial-polar

[51] Needham, SCC III: 451–4; see also Ch. 5, n. 58 (iii and vi–vii) below.
[52] Jeon, STK, p. 49.

sundial that is directly in the tradition of the Simplified Instrument and the Sun-and-Stars Time-Determining Instrument.

One such sundial is shown in Fig. 2.13. The instrument is of uncertain origin, but is probably Chinese; it is now in the collection of the Science Museum, South Kensington, London.[53] It rests on a rectangular stand equipped with three set-screws, with five-lobed flower-like grips, for levelling; an upright point (which replaces the cross-mark used with earlier sundial plummets; see Fig. 2.18) is set below the plumb-bob, for achieving a precise level. A pin-mounted compass-needle in a case affixed to the meridian partial-ring replaces the pool for a floating magnetic needle found in earlier instruments.

This sundial and a closely similar one at the Museum of the History of Science, Oxford[54] have an alidade free to rotate on the face of the disc, which is graduated only on its upper side. From the alidade, oblique threads (restored in the South Kensington instrument) attach to a tension-screw mounted on the main column, to form a triangular gnomon. This arrangement closely corresponds to that in the Sun-and-Stars Time-Determining Instrument, and similarly solves the problem posed by an equatorial sundial graduated only on its upper face: the threads of the gnomon would remain in the sunlight and could be aligned to show the time even during the six months of the year when the surface of the disc itself was in shadow. A third surviving instrument of this type, in the National Maritime Museum, Greenwich, England,[55] presumably was similarly arranged; but its alidade is missing, so that its possible use of oblique threads cannot be verified.

The South Kensington equatorial sundial and its two analogues have dials graduated in double-hours and ninety-six integral Western quarter-hours; they therefore cannot be earlier than the middle of the seventeenth century in date, and are likely to be eighteenth-century or later.

4. Small Sun-and-Stars Time-Determining Instrument (*so ilsŏng chŏngsi ŭi; hsiao jih-hsing ting-shih i* 小日星定時儀)

Sejong sillok, 77: 9a (1437):

The Sun-and-Stars Time-Determining Instrument [writes Kim Ton], as formerly constructed, was large, heavy, and inconvenient for army use, so a small form of it was

[53] Inv. no. 1981–470. [54] Cat. no. 62–104. [55] Cat. no. D.137.

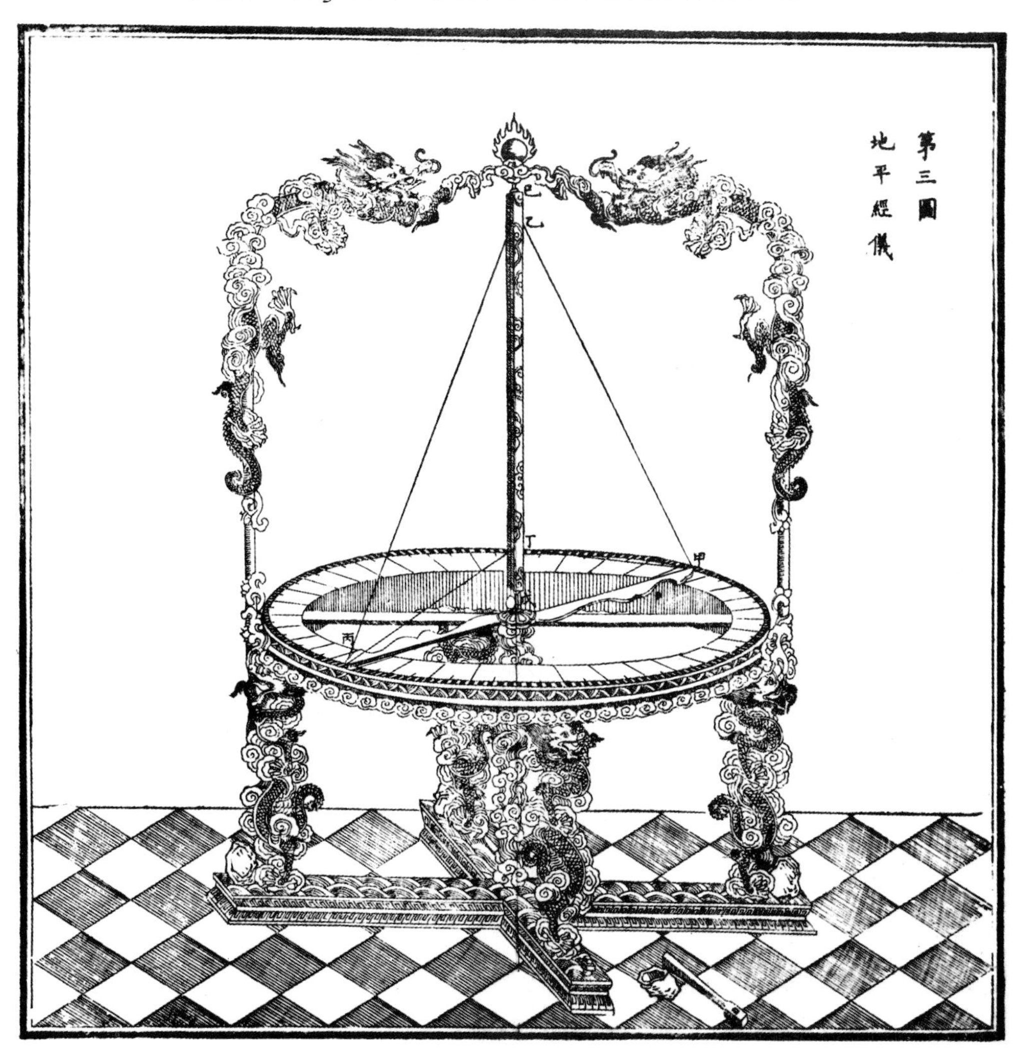

Fig. 2.12. Ferdinand Verbiest's Azimuthalem Horizontem (*ti-p'ing ching i*), mid seventeenth century; illustration from Verbiest's *Ling-t'ai i-hsiang t'u*, 1674, from the set in the Museum of the History of Science, Oxford. A fold in the original print has caused an apparent central vertical line which in the reproduction has been partly removed by retouching. Another line represents the thread suspending a plumb-bob to the left of the axis, but from Verbiest's description of the instrument (see Chapter 5, n. 58 (iii) on p. 174) and from later descriptions and photographs, it appears that the plumb-bob was in fact suspended centrally between parallel plates which form the opposite sides of a hollow square axis, and between which the diametral sight-lines pass. The mounting of the instrument is aligned to the horizon and the zenith, but its alidade and oblique sighting-threads appear to have been derived from those of Kuo Shou-ching's Simplified Instrument (Fig. 2.14) and thus to be related to those of the Sun-and-Stars Time-Determining Instrument (Figs. 2.6, 2.9) and the thread-gnomon equatorial sundial (Fig. 2.13).

made. Its design was (generally) the same as that of the previous ones, but with slight differences. The pole-determining rings (*ting-chi huan*) were omitted, so that it would be light and convenient for use.

Clepsydras are used to determine the midnight point of the day preceding the winter solstice, and observation is made of the second star of the north polar (*Pei-chi*) constellation. Then a long record-mark is made on the side of the (fixed equatorial mounting-) ring, facing north. Then three more marks are made, each shorter than the last, having a quarter-degree between each one. The first operation, done at midnight of the day preceding the winter solstice at the beginning of the year, is to align the first degree of the celestial-circumference ring with the longest mark on the (equatorial mounting-)ring. The next year it must be set against the second mark, and then against the third, and finally against the shortest of the marks. So it shifts each year. At the fifth year it is brought back to the original position.

To set the star-dial Hundred-Interval ring, the beginning is taken from the midnight of the day preceding the winter solstice at the beginning of the year. It is set opposite the zero mark of the celestial circle ring. At the first day's midnight it is set to one degree, at the second day's midnight to two degrees, at the third day's midnight to three degrees; and the same is done every year. No adjustment is made for the fractions of degrees [this having already been done by means of the celestial-circumference ring]; this is the other small difference.

The sundial Hundred-Interval ring works in just the same way as the one in the previously mentioned larger instrument.

This instrument is so similar to the full-sized version just described as to require little comment. We are not told its size, except that it was smaller than the original (the largest ring of which was 2 Chou feet in diameter). Nor does the text specify that the smaller versions were made of bronze; perhaps they were of lacquered wood, for increased lightness and portability. The pole-determining rings were omitted, but presumably something like the outer polar-constellation ring was retained, mounted on axial pillars; and there must have been similar oblique threads. Orientation was obtained by a fix on β Ur. mi. The smaller instruments were used in much the same way as the full-sized ones, except that the celestial-circumference ring itself was moved to take account of the fractional days of the four-year leap-year cycle; this meant that the daily movement of the star-dial ring was always in terms of integral degrees with respect to the celestial-circumference ring. No rule is specified for taking into account the slow drift (one *fen* in a bit over sixteen years) of β Ur. mi., as a portable instrument would have required frequent reorientation in any case.

Fig. 2.13. Equatorial sundial with double-ended alidade and triangular-gnomon thread: seventeenth-century or later (threads and plumb-bob restored).

The equatorial dial, overall diameter 90 mm, is calibrated in double-hours, each divided into four integral half-hours and subdivided into eight integral Western quarter-hours. This dates the instrument as seventeenth-century at the earliest. It is set for use at a latitude of about 35° N, and so seems more likely to be Chinese than Korean. The double-ended alidade can be reversed at noon to avoid interference between the plumb-line thread and the triangular-gnomon thread, or twisting of the two arms of the latter.

Fig. 2.14. Simplified Instrument: general view from the north-west.

This instrument, photographed by one of us (JN) at the Purple Mountain Observatory, Naking, in 1958, but formerly at Peking, is one of the exact copies of Kuo Shou-ching's instrument of 1276 made for Huang-fu Chung-ho in 1437. Other views, and an explanatory line-drawing, were published in Joseph Needham, 'The Peking Observatory in A.D. 1280 and the Development of the Equatorial Mounting', *Vistas in Astronomy*, 1955, *1*: 67–83, pp. 70–1; and still others, with a new drawing, in Needham, SCC III, Figs. 164–6. We take this opportunity to point out that each of its two equatorial alidades is diametral (see Fig. 2.15), not merely radial as previously drawn and described; and that sighting-threads (not -vanes) were used for astronomical observations made with them.

The oblique sighting-threads, which we believe led in course of time to those of the Sun-and-Stars Time-Determining Instrument (Figs. 2.6 and 2.9), of Verbiest's Azimuthalem Horizontem (Fig. 2.12), and of the thread-gnomon equatorial sundial (Fig. 2.13), do not appear in any photographs or drawings, having already long disappeared when, in Peking around 1870, a Mr W. Saunders took the photograph used by Wylie in 'The Mongol Astronomical Instruments in Peking'. According to the *Yüan shih*, Chapter 48, pp. 2b ff (repr. and tr. by Wylie, *loc. cit.*) they ran from small holes on either side of the lower end of the hollow north-polar axle to others near the end of the alidades (see Fig. 2.15) and thence back along the centre-lines of the slots in the alidades to fixing points near the equatorial axle.

5. Simplified-Instrument Platform (*kanŭi tae*; *chien-i t'ai* 簡儀臺) and

6. Simplified Instrument (*kanŭi*; *chien-i*) (Figs. 2.14–15)

Sejong sillok, 77: 9b (1437):

It was ordered by the king that Kim Ton should write a memoir on the Simplified

Instrument Platform. This said:

In the seventh year of the (Ming) Hsüan-te reign-period [= 1432], on a *jen-tzu* 壬子 day in the seventh month, His Majesty discussed the principles of calendrical and astronomical science (with his scholars and officials) in the Royal Lecture-Hall for the Exposition of the Classics. To the Academician Chŏng Inji of the Academy of Arts and Letters, he said:

'We Easterners live far across the seas, but what we do is always based on the great cultural achievements of China; only in astronomical matters have we been somewhat deficient. You, My Minister, charged with the harmonisation of calendrical computations, shall, together with the Academician Chŏng Ch'o, construct a Simplified Instrument for me.'

Therefore the officials Chŏng Ch'o and Chŏng Inji, who were conversant with the old designs (of astronomical instruments), together with Yi Ch'ŏn 李蕆 of the Privy Council, who took charge of the artisans, first made wooden models; and they checked the polar elevation, finding it to be 38.25^{d}, agreeing with the result recorded in the *Yüan shih*. Afterwards they made castings in bronze. When all was almost ready the king ordered the Financial Supervisor An Sun 安純 to build a stone observatory-platform north of the Kyŏnghoe 慶會 Pavilion in the Interior Park. The platform was 31 feet high, 47 feet long and 32 feet broad, surrounded by a little stone balustrade. On it was placed the Simplified Instrument. The True-Direction Table (*chŏngbang an*; *cheng-fang an* 正方案) was set out (*fu* 尃) to the south of this.

Kuo Shou-ching's Simplified Instrument has been described and illustrated in great detail elsewhere,[56] and so need not be treated at length here. Fig. 2.14 shows a fifteenth-century copy of Kuo's instrument in the grounds of the Purple Mountain Observatory at Nanking. Fig. 2.15 shows details of part of the instru-

[56] Needham, SCC III: 370ff and Figs. 164–6; Joseph Needham, 'Astronomy in Ancient and Medieval China', *Philosophical Transactions of the Royal Society, London*, 1974, ser. A, *276*: 67–82, Figs. 11–13; Wylie, 'Mongol Astronomical Instruments', pp. 7–13. Wylie gives a full translation of the description of the Simplified Instrument in *Yüan-shih*, 48. The history of Kuo Shou-ching's armillary sphere and Simplified Instrument and of the Ming replicas of those instruments is recounted in P'an Nai 潘鼐, 'Nan-ching ti liang-t'ai ku-tai ts'e-t'ien i-ch'i – Ming-chih hun-i ho chien-i' 南京的两台古代测天仪器～明制浑仪和简仪 (Two ancient astronomical observational instruments at Nanking – the armillary sphere and Simplified Instrument made during the Ming period), *Wen-wu*, 1975.7: 84–9.

I Shih-t'ung 伊世同 'Cheng-fang an k'ao', *Wen-wu* 1982.1: 76–7,82 shows that the True-Direction Table (*cheng-fang an* 正方案) survives, in part, as a square plate which lies horizontally below the S pole of the Simplified Instrument. The plate formerly carried a vertical gnomon surrounded by nineteen concentric engraved circles, and was used for determining (or demonstrating) the true meridian by the method of bisecting straight lines drawn between points of equal gnomon-shadow radius, as described in the *Yuan Shih*, ch. 48, 7b–8b. In the seventeenth century it was re-engraved for use as a horizontal sundial, and was equipped with a polar-pointing gnomon. (In 1958 a gnomon of this form was seen in situ – JN.)

ment, including the bronze mobile declination split-ring, which carries a sighting-bar, and the mobile equatorial ring, with its alidades – the ancestor of the Sun-and-Stars Time-Determining Instruments described above. The Simplified Instrument was essentially an observational armillary sphere, dissected so that its rings were arrayed separately rather than in a concentric nest. Its equatorial-polar alignment (derived from Chinese armillary spheres) anticipated by three centuries the mounting arrangement of modern astronomical instruments as they developed in Europe from the time of Tycho Brahe onwards. Kuo's instrument seems to have been influenced by the *torquetum* of the Islamic tradition of astronomical instrumentation.[57]

Kim Ton's account of the casting, in 1432, of King Sejong's Simplified Instrument is very brief. A wooden prototype was made, the polar elevation was checked, and the instrument was cast in bronze. Kim seems to have taken for granted that not only the court astronomers, but also the future readers of his memoir, would know exactly what a Simplified Instrument was, what it looked like, and how it was used; thus he felt no need to go into such matters. It seems unlikely that the writer of an official memoir would have treated so momentous an event as the casting of a great instrument so cursorily had he not been able to assume that all concerned would understand its import on the basis of a few lines. His apparent confidence that everyone at King Sejong's court knew all about Simplified Instruments leads us to wonder whether the Korean court might have possessed one even prior to the great programme of instrument-making inaugurated in 1432.

We have not been able to find any earlier reference to the casting or importation of a Simplified Instrument in Korea, but such events could still have occurred. Kuo Shou-ching's Shou-shih calendar of 1280 was adopted by the Koryŏ court in 1281; complete descriptions of the instruments of Kuo's observatory would almost certainly have accompanied the new calendar to the Korean court. In 1308 the Bureau of Astronomy (Sŏun Kwan) was created by the amalgamation of the Office of History and Portent Astrology (T'aesaguk, Ta-shih Chü 大史局) and the Observatory (Sach'ŏn Tae, Ssu-t'ien T'ai 司天臺); Jeon speculates that 'this change was presumably accompanied by a large-scale reconstruction and renovation of the observatory facilities'.[58] It seems not unlikely that a Simplified Instrument would have been made part of the observatory's

[57] Needham, SCC III: 370–2.   [58] Jeon, STK, p. 37.

Fig. 2.15. Simplified Instrument: close-up view from the north-west.

This view shows the fixed diurnal ring (the *Yüan shih* states the diameter to be 6 feet 4 inches; we are unsure of the exact dimensions of the 'foot' in this case, but the diameter must be close to 2 m) upon which the mobile equatorial ring is supported by four roller-bearings. Two pointed and slotted alidades are pivoted to rotate above these rings. In the original photograph (JN, 1958) small holes for the sighting-threads (see Fig. 2.14 caption) are visible just beyond the outer ends of the slots; Wylie reports the appearance of a similar hole on one side of the polar axle in Saunders's photograph.

Provision of two alidades facilitated simultaneous determinations of the positions of two celestial objects. Readings on the scale-rings were indicated by the pointed ends of the alidades, rather than by the sighting-thread itself as in the Sun-and-Stars Time-Determining Instrument (Fig. 2.9).

The photograph also shows the mobile solar-declination ring (diameter 6 Yüan feet), with its cross-struts and centrally pivoted alidade with sighting vanes, and also an inclined pointer for scale-readings on the equatorial rings. The solar-declination ring and the two equatorial alidades with sighting-threads must have been moved out of each other's way as necessary during use.

equipment at that time. A further opportunity for constructing a Simplified Instrument for the Koryŏ observatory would have come after 1370, when the *Yüan shih* (which had been compiled with unusual speed) was completed. The 'Treatise on Astronomy' (*T'ien-wen chih* 天文志; *Yüan-shih*, 48) of the Yüan history contains a detailed description of the Simplified Instrument. The ***kanŭi*** with which King Sejong's officials were apparently so familiar may thus have been one possessed by the court in the days of the Koryŏ Kingdom. On the other hand, there is no indication that a Simplified Instrument existed in Korea at the time when the new one was commissioned in 1432; if there had been an earlier one, perhaps it was damaged or destroyed at the time of the fall of the Koryŏ rulers.

In any case, the casting of the new Simplified Instrument marked the first step in King Sejong's ambitious programme of re-equipping the Royal Observatory. It was an event of sufficient importance to justify the building, apparently at considerable expense, of a new observatory platform as well. The Simplified-Instrument Platform not only housed the new *kanŭi* itself, but many of the king's other new instruments were also to be placed on or near it. When it was fully equipped, it must have resembled closely Kuo Shou-ching's own observatory from the 1270s; unfortunately, the appearance of that observatory too can only be surmised from literary records. Its instruments have long been scattered, and some were melted down.[59] Fifteenth-century copies of some of Kuo's original instruments are now (or were in 1958) at the Purple Mountain Observatory in Nanking, but their arrangement there bears only a conjectural resemblance to that of the original observatory in Peking. The best reconstruction of the layout of the Yüan observatory is that of Yamada Keiji:[60] we have reproduced this as Fig. 1.3 above. Jeon illustrates remains of a seventeenth-century observatory, which are similar to surviving remains of the almost identical fifteenth-century observatory in Seoul; as well as a detail from a painting of another 17th-century observatory platform with several instruments in place.[61] One may also compare illustrations[62] of the seventeenth-century Jesuit observatory in Peking to gain some impression of what King Sejong's observatory may have looked like. Whatever its actual layout, the *kanŭi tae* with its full set of new instruments must have been a place of considerable grandeur.

7. Small Simplified Instrument (*so kanŭi*; *hsiao chien-i* 小簡儀)

Sejong sillok, 77: 9a (1437):

The Preface for the Small Simplified Instrument was written by Chŏng Ch'o,

[59] Needham, SCC III: 452.

[60] Yamada Keiji 山田慶児, *Jujireki no michi* 授時暦の道 (The principles of the Shou-shih calendrical system) (Tokyo: Mizusu Shobo, 1980).

[61] Jeon, STK, Figs. 1.6–7.

[62] Needham, SCC III: Figs. 190–1 (following p. 450). See also Ch. 5 n. 58 (vi), (vii), on p. 174 below.

Intendant of Education of the Bureau of Arts and Letters. It said:

When T'ang 湯 and Yao were ruling the world they first ordered Hsi and Ho to calibrate the sundial. Thenceforward every dynasty has made instruments. By the time of the Yüan Dynasty the designs approximated to today's.

In the sixteenth year of His Majesty's reign (1434), in the autumn, He ordered Yi Ch'ŏn, Chŏng Ch'o, and Chŏng Inji, among others, to construct a small version of the Simplified Instrument, following the old methods but using a new type of base. Made of pure bronze (this base) is provided with water-grooves to fix the level and to adjust the north–south alignment [by means of a floating compass needle?].

(With regard to the instrument itself), one of the equatorial rings is graduated throughout the circle with degrees and fractions of degrees, and it can turn through east and west to take measurements of (the movements of) the Seven Regulators [i.e. sun, moon, and five visible planets], and to read off the positions of the stars and constellations north and south of the equator with regard to the *hsiu* [Lunar Lodge extents in] degrees and fractions of degrees [i.e. right ascension].

The Hundred-Interval ring is inside the (mobile) equatorial ring, and it is marked with the Twelve Double-hours and the Hundred Intervals (*k'o*). Thus by day the sun's position can be known, while by night the culminations of the stars can be ascertained. The Ring of the Four Displacements (*ssu-yu huan* 四游環) [the mobile declination ring] carries a sighting alidade, and can turn east and west while at the same time the sighting alidade can move north and south, so that it can be used for all positional measures.

The three rings are upheld by pillars in a slanting position so that the mobile declination ring is set (with its diameter) in the polar axis; similarly the equatorial ring is set so as to be in the same plane as the belly of the heavens. Thus when the mobile declination ring is set straight upwards the four cardinal points are established, and when (the equatorial ring) turns round about the Hundred-Interval ring the coordinates of the night are found out.

When the work was finished, authority was sought for the carving of an inscription. Your Majesty asked His servant (Chŏng) Ch'o for the inscription, which he now accordingly presents as follows:

The Tao of Heaven is unforced activity; the best instruments are the simplest. The old Simplified Instrument was bulky and clumsy in its framework and base, while the new one, though having the same uses, has been made conveniently portable. So it is even simpler than the old Simplified Instrument.

Kim Ton, in his memoir on the Simplied-Instrument Platform, adds the following (*Sejong sillok*, 77: 10a):

Although the Simplified Instrument is simpler than the armillary sphere, it is difficult to turn and use, so two Small Simplified Instruments were made. Although (even more) simple, their use is the same as that of the Simplified Instrument. One of them was placed to the west of the Thousand Autumns Hall, while the other was given to the Bureau of Astronomy.

The Simplified Instrument of 1432 was presumably a close copy of Kuo Shou-ching's instrument, and must have been large and massive; the instrument (Fig. 2.14) that still survives at (or from) the Purple Mountain in Nanking (a fifteenth-century copy of Kuo's original) has declination and equatorial rings that span nearly two meters. Two years after casting a full-sized Simplified Instrument, King Sejong's astronomers decided to make a less unwieldy version. We are not told its dimensions, but in order to be 'conveniently portable' it must have been very much smaller. It was provided with an improved water-level base, the reservoir of which could also have held a floating compass-needle. The vertical altitude-measurement ring of the full-scale instrument[63] seems to have been omitted in the smaller ones; it is not mentioned in Chŏng Ch'o's Preface.

The description here of the fixed equatorial Hundred-Interval ring mounted within a mobile celestial-circumference ring makes clear once again that the Sun-and-Stars Time-Determining Instrument described above was derived from the Simplified Instrument. Neither the full-scale Simplified Instrument nor the two smaller versions of 1434 survived the Hideyoshi invasions of 1592–8, and no illustrations of them are known.

8. Measuring-Scale and Gnomon (*kyup'yo*; *kuei-piao* 圭表), with a Shadow-Definer (*yŏngbu*; *ying-fu* 影符) (Fig. 2.16)

Kim Ton's memoir on the Simplified-Instrument Platform continues:

Sejong sillok, 77: 9b (1437):

West of the Observatory Platform there was set up a bronze gnomon five times 8 feet [40 (Chou) feet] in height. Its shadow-measuring scale was made of blue-green stone [marble] with graduations on it in tens-of-feet, feet, inches, and *fen* [tenths-of-inches]. A Shadow-Definer was used to focus the shadow cast by the sun's centre, measuring the (gnomon) shadow length precisely at its longest and shortest in winter and summer.

This instrument evidently was based on Kuo Shou-ching's 40-foot gnomon, of

[63] Ibid. 371, Fig. 166, letter 'n'.

which the following description is given in the *Yüan shih* (48: 8b):[64]

The graduated scale is made of stone 128 feet long, 4.5 feet wide, and 1.4 feet thick; its base is 2.6 feet high. Circular basins are excavated at the north and south ends, each 1.5 feet in diameter and 2 inches deep. From 1.0-foot distance from the gnomon for 120 feet a central strip of 4 inches wide is marked off, 1 inch on each side of which is divided into feet, inches, and tenths, extending to the north end. One inch from the edge are water-channels, 1 inch deep, connected with the reservoirs at the ends, for the purpose of levelling.

The gnomon is 50 feet long, 2.4 feet wide, and 1.2 feet thick [of bronze, or perhaps of brass], and is fixed in the stone base at the south end of the graduated scale. Inserted to a depth of 14 feet in the earth, it rises 36 feet above the scale. At the top the gnomon divides into two dragons which sustain a cross-bar (*heng-liang* 横梁). From the centre of the cross-bar the measurement to the top of the gnomon is 4 feet, and hence to the top of the scale it is 40 feet. The cross bar is 6 feet long and 3 inches in diameter, and carries a water-channel on the top for the purpose of levelling. At its two ends and in the centre are transverse holes, $\frac{1}{5}$ inch in diameter, through which are inserted rods 5 inches long, having plumb-lines attached to them so that the correct position can be ascertained and lateral deflection prevented.

When a gnomon is short, the divisions on the scale have to be close together and very small, and most of the smaller divisions below feet and inches are difficult to determine. When a gnomon is long, the graduations are easier to read, but the inconvenience then is that the shadow is light and ill defined, making it difficult to get an exact result. In former times, observers sought to ascertain the real point by using a sighting-tube, or a small gnomon, or a wooden ring; all devices for easier reading of the shadow-mark on the scale. But now with a 40-foot gnomon, 5 inches of the graduation scale corresponds to what was only 1 inch previously [on the shadow-scale for the 8-foot gnomon] and the smaller subdivisions are easier to distinguish.

The Shadow-Definer (*ying-fu*) is made of a leaf of copper 2 inches wide and 4 inches long, in the middle of which is pierced a pin-hole. It has a square supporting framework, and is mounted on a pivot (*chi chu* 機軸) so that it can be turned at any angle, such as high to the north and low to the south [i.e. at right angles to the incident sunlight]. The instrument is moved back and forth until it reaches the middle of the (shadow of the) cross-bar, which is not too well defined, and when the pin-hole is first seen to meet the light, one receives an image no bigger than a rice-grain in which the cross-bar can be noted indistinctly in the middle. With the old methods, using the simple summit of the gnomon, what was projected was the upper edge of the solar disc. But with this method

[64] Quoted from Needham, SCC III: 298–9, slightly modified.

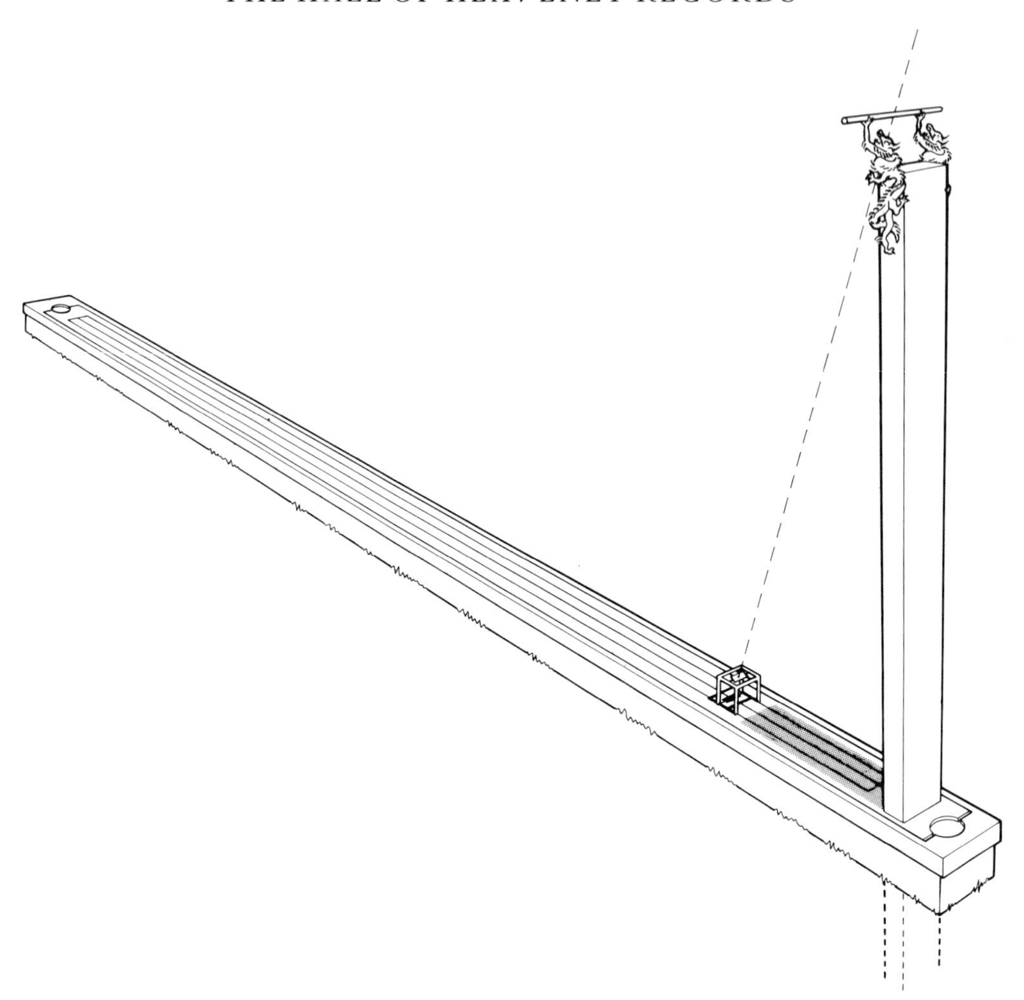

Fig. 2.16. Measuring-Scale with 40-Foot (9.8 m) gnomon, and Shadow-Definer with 2-inch by 4-inch (49 × 98 mm) vane: reconstruction drawing.

This reconstruction drawing is based on the description (see p. 71) of the 40-foot gnomon erected in Peking by Kuo Shou-ching and on Ferdinand Verbiest's drawings (Chapter 5, n. 58 (v) below) of an 8-foot gnomon of similar design. We believe that King Sejong's 40-foot gnomon was copied from Kuo's instrument; but it is not certain that the Korean instrument was as elaborately ornamented as the Chinese original was.

In the close-up detail view of the Shadow-Definer, the tiny 'optical' image of the sun (with part of the gnomon cross-bar), which is formed by the pin-hole in the vane, has been greatly exaggerated in size to make it visible in the reproduction.

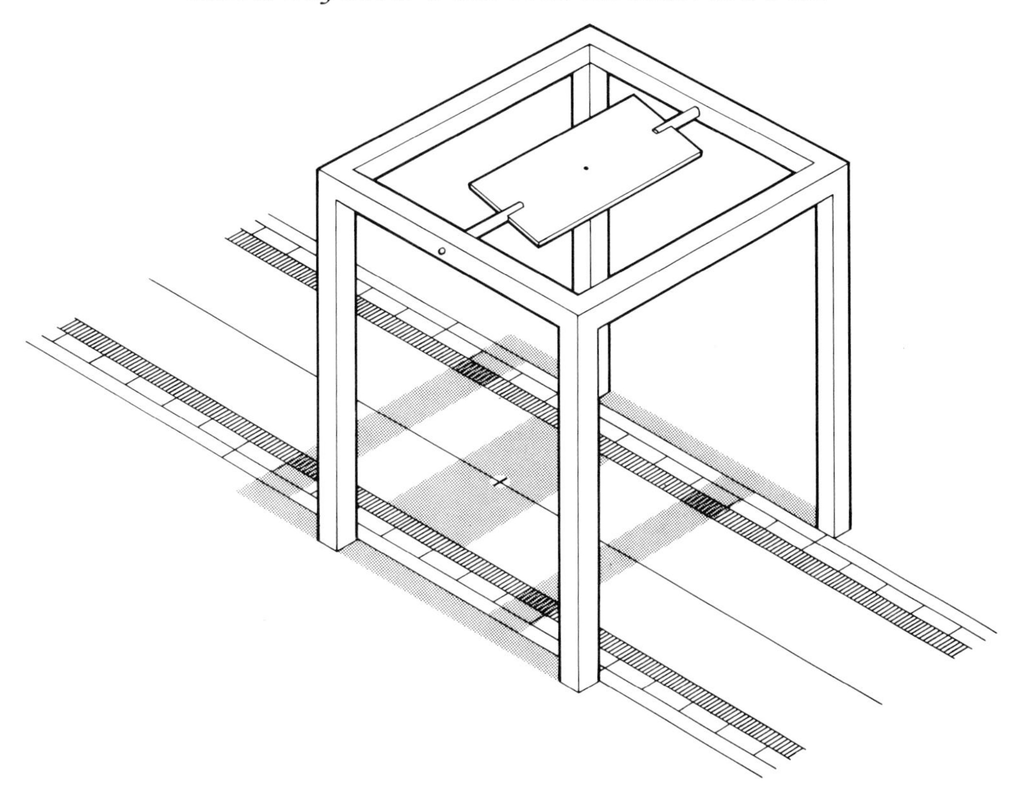

Fig. 2.16. (detail).

one can obtain, by means of the cross-bar, the rays from the centre of the disc without any error.

Two of Kuo Shou-ching's gnomons were erected, at Peking and at Yang-ch'eng 陽城 (modern Teng-feng 登封 County), near Loyang. The one at Yang-ch'eng was supplied with a splendid semi-pyramidal tower enclosing the gnomon;[65] the tower and its shadow-scale remain to this day, although the gnomon itself is missing. King Sejong's gnomon was apparently not provided with such a tower, but rather was free-standing like the one at Peking. Figure 2.16 shows how it may have appeared.

[65] Needham, SCC III: 296–9 and Figs. 115–17; also Chang Chia-t'ai 张家泰, 'Teng-feng kuan-hsing-t'ai ho Yüan-ch'u t'ien-wen kuan-ts'e ti ch'eng-chiu' 登封观星台和元初天文观测的成就 (On the observation tower at Teng-feng and the achievements of astronomical observation in the early Yüan period), *K'ao-ku* 考古, 1976.2: 95–102, repr. in *Chung-kuo t'ien-wen-hsüeh shih wen-chi* 中国天文学史文集 (Collected essays in the history of Chinese astronomy) (Peking, 1978), pp. 229–41.

Apart from its great size, the most interesting feature of this gnomon and shadow-scale is its Shadow-Definer, invented by Kuo Shou-ching. Previously, the great barrier to the use of large gnomons was the phenomenon of the penumbra, which caused the edge of the shadow to become increasingly indistinct the longer the gnomon became, regardless of how sharp one made the top edge of the gnomon itself. The Shadow-Definer (Fig. 2.16 detail), moved along the shadow-scale and tilted to face the incident sunlight, used the principle of the pin-hole to focus, like a lens, the image of the cross-bar.[66]

Kim Ton specifies the height of this gnomon as being 'five times 8 feet'. From the Chou period onwards, 8 feet had been the standard height of Chinese gnomons;[67] Kim's statement thus would serve as a reminder that the results of shadow-length measurements obtained with the new gnomon could be compared with those given in earlier astronomical treatises by simply multiplying the earlier figures by five.[68]

9. Armillary sphere (*honŭi*; *hun-i* 渾儀) and

10. Celestial globe (*honsang*; *hun-hsiang* 渾象)

Kim Ton's memorial continues,

Sejong sillok, 77: 9b (1437):

West of the gnomon they set up a small pavilion to house the armillary sphere and the celestial globe, the sphere being to the east and the globe to the west. The design of armillary spheres has not been the same all through the ages, so in this case (the makers) relied on the writings of Mr Wu 吳氏 and used lacquered wood for the rings of the sphere. The celestial globe was made of lacquered cloth (on a framework), round as a crossbow-bullet, having a circumference of 10.86 feet.[69] Coordinates were marked on it

[66] Needham, SCC III: 299.

[67] *Ibid.* 286; Nakayama, *History of Japanese Astronomy*, pp. 25–9.

[68] Cf. Jeon, STK, p. 38.

[69] The apparent precision of this figure is probably misleading. For example, it could indicate that the *diameter* of the celestial globe was 3.4 (Chou) feet, and that 3.2 was used as an approximation of π in calculating its circumference; this would yield a figure of 10.88 feet rather than 10.86 feet as given. In everyday mathematics in China, 3.2 was often used as an approximation of π, even though a quite accurate calculation of the value of π had been performed as early as the fifth century C.E. (see Needham, SCC III: 99–102). Alternatively, if the diameter of the globe were 3.4 feet and the value of π were taken as 3.1416, the calculated circumference would be approximately 10.681 feet; thus 10.86 may be a misprint for 10.68. On the other hand, the circumference could have been calculated by specifying a nominal 0.3 *tsun* per *tu*; in that case, the figure 10.86 might be a misprint for 10.96 (i.e. 365.25×0.03). It is also possible that the figure 10.86 is quoted with erroneous 'precision' from an empirical measurement.

in celestial degrees; the equator was in the middle with the ecliptic crossing it at an angle of a little under 24 degrees. All over the cloth surface were marked the constellations north and south of the equator.

In one day it made one revolution with an addition of one degree. There was a model of the sun which was moved by means of a silk thread along the ecliptic one degree each day, exactly corresponding to the motion of (the real sun in) the heavens. And all the ingenious mechanism of its drive by trickling water was hidden so that it was invisible.

We wish that Kim Ton had provided more information about the drive-mechanism of this globe and sphere. Such important instruments, we assume, must have been the subject of commemorative inscriptions, prefaces, and memoirs; the Veritable Records do not quote them. On the basis of Kim Ton's brief reference to a hidden water-operated mechanism, it seems reasonable to think that the armillary sphere and celestial globe were driven automatically, employing the Sino-Arabic technology that had been adapted for use in the Striking Clepsydra. It is probable also that the two instruments were rotated by a single mechanism.

As we note in more detail in the next chapter, the globe and sphere were repaired in 1526 and replaced in 1549; a further replacement was commissioned in 1601 after the catastrophe of the Hideyoshi invasions. The second replacement was described as a *sŏn'gi okhyŏng*, by that time the standard term for an armillary sphere, often clockwork-operated. In any case, it is certain that Yi Minch'ŏl's 1669 armillary clock was operated by a clepsydral mechanism, and there is no reason why King Sejong's globe and sphere should not have been also, as a part of a continuous tradition. King Sejong's Jade Clepsydra (no. 11, below) incorporated still other features derived from Arabic clepsydral technology, so we know that the full range of such mechanisms was available to the royal engineers.

Beyond this, we can say nothing about the drive-mechanism of King Sejong's sphere and globe. From a position of extreme scepticism one might even doubt that it was mechanised at all; the reference to 'mechanism . . . of trickling water' etc. might simply be traditional verbiage such as was applied to armillary spheres from the time of Chang Heng onwards. But we are inclined to reject this position.

The sphere and globe described by Kim Ton were made of lacquered wood and cloth; their resulting lightness would have made them easy to rotate. Jeon notes that another chapter in the Veritable Records of King Sejong (60: 38b) records the casting, in 1433, of additional bronze armillary spheres under the supervision of the same group of officials whose work we are examining in this chapter.[70] There is no indication that any of these bronze armillary spheres were mechanically operated in any way; they were most likely simply copies (perhaps on a smaller scale?) of Kuo Shou-ching's great sphere.[71] Another Yi-period armillary sphere, illustrated by Rufus,[72] is probably of later manufacture.

We are unable to identify with complete assurance the Mr Wu on the basis of whose designs the lacquer sphere and globe were built. One possibility is Wu Chao-su 吳昭素, author of the calendar of 981, who was mentioned in the *Sung shih* in connection with an armillary sphere built by Han Hsien-fu 韓顯符.[73]

At this point in his memoir on the Simplified Instrument Platform, Kim Ton says, 'These five instruments [i.e. the Sun-and-Stars Time-Determining Instrument, the Simplified Instrument, the gnomon, the armillary sphere, and the celestial globe] are all described in detail in the ancient texts.' Presumably he means that these instruments were derived from the work of Kuo Shou-ching, who, having lived 150 years earlier than Kim, could reasonably be described as 'ancient'.

11. The Jade Clepsydra (*ongnu*; *yü-lou* 玉漏)

After briefly describing the Striking Clepsydra housed in the Porugak (no. 1, above), Kim says:

Sejong sillok, 77: 10a (1437):

West of the Hall of a Thousand Autumns was built a small pavilion called the Hall of Respectful Veneration (Hŭmgyŏnggak, Ch'in-ching Ko 欽敬閣). There, paper was pasted together (over a framework) to make a mountain more than seven (Chou) feet in height, within the pavilion. Inside there was set up a (jack-)mechanism wheel (*chi lun* 機輪) which used water from a jade clepsydra (vessel) to (operate its) striking (apparatus).

There were five(-coloured) clouds surrounding the sun, which rose and set, while at the same time a jade girl (immortal) struck the double-hours upon a bell with a wooden mallet. There were four jacks in the form of warriors, arranged to face each other, and twelve jacks which each in turn rose up the mountain and retired again as their turning (mechanism) brought them round. On the four sides of the mountain there was scenery displaying the works and days of the four seasons according to the descriptions in the

[70] Jeon, STK, p. 66. [71] Needham, SCC III: Figs. 156, 163.
[72] Rufus, 'Astronomy in Korea', Fig. 32, top right.
[73] Needham, Wang, and Price, HC, pp. 68–70.

Odes of Pin,[74] to remind the onlooker of the hard toil of the people for the production of food and clothing.

There was also set up an inclining vessel (*i ch'i* 欹器) to receive the overflow water of the clepsydra's constant-level tank, and so to investigate the Tao of the heavens by the principle of fullness and emptiness.

Chapter 2 of the *Chŭngbo munhŏn pigo* (pp. 30a ff) provides further information on this instrument:

(In the Hŭmgyŏnggak) there was set up a Jade Clepsydra with a wheel which was turned by the action of water ... There was a sun made of gold [i.e. gilded?], about the size of a crossbow-bullet (*tan-wan* 彈丸), which revolved around (the paper mountain) once a day ...

There was also a platform, on top of which there was an inclining vessel, and north of the platform there was a (figure of an) official holding up a metal vase from which water poured forth; this was the excess of the clepsydra water which poured down continually without ceasing. (The vessel) when empty lay on its side, when half full stood upright, and when full overturned ...

Everything was like the traditional designs. For all the motions there was apparatus moving by itself and striking by itself without the need of any person to be in the pavilion.

The most complete description of the Jade Clepsydra, however, is found in a separate 'Memoir on the Hall of Respectful Veneration' (*Hŭmgyŏnggakki* 欽敬閣記), written by Kim Ton in 1438, and quoted in Chapter 80 of the Veritable Records of King Sejong:

Sejong sillok, 80: 5a–6a[75] (1438):

A pasted-paper mountain about seven feet high is set up in the Hŭmgyŏnggak. Inside are set up the Jade Clepsydra (jack-)mechanism wheels, made to rotate by means of trickling water. A golden image of the sun, the size of a crossbow-bullet, is provided to move across the middle of the mountain, surrounded by multicoloured clouds. The sun makes its daily revolution, appearing at the rim of the mountain at dawn and hiding behind it at dusk. The declination of the sun varies depending upon the polar distance, while its rising and setting follow that of the real sun for each of the Fortnightly Periods.

[74] 'Pin feng' 豳風, one of the sections of the *Shih ching* 詩經, *c*. 700 B.C.E. or earlier. The reference here is especially to the poem 'In the Seventh Month', Mao ode no. 154; see Bernhard Karlgren, *The Book of Odes* (Stockholm: Museum of Far Eastern Antiquities, 1950), pp. 97–9; also Arthur Waley, *The Book of Songs* (London: Allen and Unwin, 1937), pp. 164–7 (this poem is no. 159 in Waley's revised numbering system).

[75] Tr. Jeon, STK, pp. 60–1, modified.

Below the sun-model stand four jade female immortals placed at the four cardinal points, each with a golden bell in hand. At the beginning and middle of each of the three morning double-hours ranging from *yin* 寅 to *ch'en* 辰, the immortal in the east rings the bell, followed by the immortal to the west for the beginnings and middles of the next three double-hours, and similarly by those to the north and the south by turns.

At the four cardinal points on the ground stand the Four Gods (*ssu-shen* 四神),[76] each facing the central mountain. The first of the Four Gods, the Blue-Green Dragon, faces north at the double hour of *yin* 寅, south at *mao* 卯, and west at *ssu* 巳. Similarly the Red Bird faces first east (at *yin*), and so on. At the southern foot of the mountain stands a high platform, where an 'hour-jack' (*ssu-ch'en* 司辰) stands with his back turned to the mountain, while three warriors, all in armour, are arranged in such a way that the first one, carrying an iron hammer, stands in the east facing west; the second one, carrying a drumstick, stands in the west facing east; and the third one, carrying a gong-stick, also stands in the west facing east. At each double-hour, the hour-jack turns to face the bell-striker; (the other jacks) strike their respective instruments at each night-watch. On the level ground are the Twelve Gods (*shih-erh shen* 十二神) occupying their respective positions, and behind them are holes. At the hour of *tzu* 子 the hole behind the Rat opens; an immortal with a time-tablet (*shih-p'ai* 時牌) comes out and the Rat stands still. At the double-hour of *ch'ou* 丑 there is a similar performance at the place where the Ox is standing, and so on through the successive double-hours.

South of these is located another platform carrying an inclining vessel (*i ch'i*), which lies on its side when empty, stands upright if half full with water, and falls over again if filled to the brim. All of these operations are performed completely automatically without any help from anyone. Around the mountain is an enclosure with paintings of rural scenery in the four seasons, and wood carvings of men, birds, and plants displaying the labours undertaken by the people in the different seasons.

Like the Striking Clepsydra, this instrument is a direct descendant of Yüan Shun-ti's palace clepsydra of the 1350s; it also seems reminiscent of the Precious Mountain Clepsydra (Pao-shan Lou 寶山漏), built by Kuo Shou-ching for Kubilai Khan in Shangtu in 1262.[77] Both of these timekeepers, similar fantasias of scenery, jade immortals, promenading officials, etc., might have had falling balls as their main power-transmitting devices, although surviving descriptions make

[76] The Four Gods – the Blue-Green Dragon of the east, the Red Bird of the south, the White Tiger of the west, and the Dark Warrior (a paired turtle and snake) of the north – had symbolised the cardinal directions in Chinese cosmology at least since the mid-Chou period. See John S. Major, 'New Light on the Dark Warrior', in N. J. Girardot and J. S. Major, eds., *Myth and Symbol in Chinese Tradition* (Boulder, Colo.: *Journal of Chinese Religions* Symposium volume, 1985).

[77] *Yüan-shih*, 5: 2a; Needham, Wang, and Price, HC, p. 135, n. 6.

no mention of them.[78] The descriptions of the Jade Clepsydra quoted above also make no mention of falling balls, although we assume that, like the Striking Clepsydra, at least some of its effects were operated in that way. On the other hand, its main power-train might have been worked by an anaphoric device, i.e. an axle or drum rotated by means of a weighted cord attached to the clepsydra float.

All of the descriptions make prominent mention of the 'inclining vessel' (*i ch'i*) that filled and emptied automatically. This must have been a pivoted container arranged so that its centre of gravity changed according to the amount of water it held; it was conceptually analogous to the water-operated bamboo clappers that are a familiar feature of Japanese gardens. The inclining vessel is specified as having been filled by the overflow water from the clepsydra's constant-level tank, and therefore it could not (because its water supply was unregulated) have served a timekeeping function. It was essentially a showman's device to impress the onlookers; the effect depended on the classical associations of such vessels. The philosopher Hsün-tzu, writing in the third century B.C.E., described an 'overturning vessel' that was placed on the right-hand side of the throne in the ancestral temple of the ancient state of Lu; it was understood to serve as a reminder of the virtue of moderation.[79] In view of the conventionally Taoist character of the Jade Clepsydra's imagery, the device might have been intended also to recall to mind the philosopher Chuang-tzu, whose teachings were described as 'goblet words' (*chih yen* 卮言) that 'upset themselves when full and righted themselves when empty'.[80]

The Jade Clepsydra had three main time-annunciating mechanisms:

1. Four female immortals stood on peaks of the mountain, in the four cardinal directions. They held bells that were struck with wooden mallets at the double-hours.

2. A main hour-jack stood on a platform in front of the mountain, carrying a bell, a drum, and a gong; presumably these were hung on a rack held in the jack's outstretched hands. In the east and somewhat to the front of that jack stood another holding a bell-mallet; to the west stood two jacks, one with a drumstick, the other with a gong-stick. At the beginnings and middles of the double-hours,

[78] See n. 30 above.

[79] *Hsün-tzu* 荀子, 28: 1a, tr. and discussed in Needham, SCC IV.1: 34–5.

[80] *Chuang-tzu*, Chs. 27, 33. Burton Watson, *Chuang-tzu: Complete Works* (New York: Columbia University Press, 1968), pp. 303, 373. See also Needham, Wang, and Price, HC, pp. 84–9, 94.

the jack would turn to the mallet-holding jack to the east, which would strike a signal on the bell. At the beginning of each night-watch, the main jack would turn towards the two figures to the west, which would strike the drum and the gong. At the next four *tien* (fifths of night-watches) only the gong jack would strike.

3. Stationary double-hour gods (the Rat, Ox, etc.) occupied positions in front of slots. At each double-hour, a placard-bearing jack would rise into position behind the appropriate double-hour god.

In addition, a sun-model revolved around the mountain once each day. Its path was adjusted periodically so that it appeared and disappeared approximately at sunrise and sunset throughout the year.

One can easily see how these timekeeping functions could have been performed by means of falling balls, as in the Striking Clepsydra. The motion of the sun could have been taken directly from an axle driven by the clepsydrally released balls, and this was presumably also done in the case of the celestial globe described above (no. 10, pp. 74–6).

The descriptions of the Jade Clepsydra claim that all of the effects of the mechanism were completely automatic, but that cannot have been true – the clepsydra vessels would have had to be filled and emptied (presumably morning and evening daily), while ball-racks operated by float-rods (or whatever else was used to time the seasonally variable night-watches) would have had to be changed manually at fifteen-day intervals. Probably the height and/or angle of the sun's path would have been adjusted manually at the same intervals.

The mountain itself was designed in the Chinese cosmological tradition of Mt K'un-lun (=Mt Meru, the celestial *axis mundi*). With its female immortals standing on crags, its multicoloured clouds, etc., it must have been like a large version of the Han 'hill censers' and their descendants in bronze and porcelain.[81]

12. Scaphe sundials (*angbu ilgu*; *yang-fu jih-kuei* 仰釜日晷) (Fig. 2.17)

After making the passing reference to Small Simplified Instruments that we quoted in connection with no. 7 above, Kim Ton goes on to say,

[81] For a discussion of the symbolism of these objects, see Homer H. Dubs, 'Han Hill Censers', in E. Glahn and S. Egerod, eds., *Studia Serica Bernhard Karlgren Dedicata* (Copenhagen: Munksgaard, 1959), pp. 259–64; also Needham, SCC III: 580–1. See also the splendid example illustrated in Wen Fong, ed., *The Great Bronze Age of China* (New York: Metropolitan Museum of Art, 1980), Pl. 95 and Fig. 115.

Sejong sillok, 77: 10a (1437):

> As ordinary people were rather ignorant about double-hours and 'intervals' (*k'o*), two hemispherical scaphe sundials were made, with the (twelve) double-hour gods inscribed inside them. Thus the common people could tell the time. One was set up halfway across the Benevolent Government Bridge, and the other in South Royal Ancestral Temple Street.

Kim goes on to say that because such sundials were useless for telling the time at night, Sun-and-Stars Time-Determining Instruments were made.

Scaphe sundials were among the instruments made by Kuo Shou-ching; we have no information about them in East Asia from any earlier period, though hemispherical sundials had long been known in the West.[82] In their usual form, they were graduated with a reticular scale inside the hemispherical scaphe; Fig. 2.17 shows a typical Korean example from the seventeenth century. By means of this scale, the scaphe sundial gives both the time of day, by hour-circle lines, and the season (in Fortnightly Periods), by solar-declination lines.[83] Although Kim Ton does not mention them, scaphe sundials of this typical sort must have been made for use within the palace, and no doubt for the Royal Observatory as well. This we infer from his statement that 'since (sundials) were of no use in telling the time at night, *ilsŏng chŏngsi ŭi* were made'. As the latter were not for public use, the comparison made in this statement must refer to scaphe sundials for palace use. The author seems to have taken them for granted.

The unusual feature of the public scaphe sundials that Kim Ton does single out for special mention is that they were not graduated with a reticular scale for the hours and intervals (*k'o*) and for the Fortnightly Periods, but with representations of the Rat, the Ox, the Tiger, and the other double-hour gods. These had been used by ordinary folk in East Asia to designate diurnal time at least since the Chou period,[84] and would have been comprehensible even to the most horologically unsophisticated passer-by. We may assume that Kim Ton mentioned this modification of a sophisticated sundial for the use of an illiterate public not only because it was innovative, but also as a conspicuous tribute to the public-spirited beneficence of his ruler.

[82] Needham, SCC III: 301.

[83] See also Needham, SCC III: Fig. 123a, and Jeon, STK, Fig. 1.11. Our present study of scaphe sundials supersedes SCC III: 302, n. b, which erroneously suggests that scaphes depend on shadow length, not direction.

[84] Peter A. Boodberg, 'Chinese Zoographic Names as Chronograms', *Harvard Journal of Asiatic Studies*, 1940, 5: 128–36.

The characteristic shape of the pole-pointing gnomon of scaphe sundials in the East Asian tradition – a double-ogee cone on a Greek-cross base – deserves mention. One might speculate that it is of Islamic origin; on the other hand, it could also be indigenous to East Asia. The shape and polar orientation seem to echo Mt K'un-lun, the cosmological polar pivot of the universe; while the tines of the double-ogee cone could be derived from the claws of the *vajra* ('thunderbolt') ritual implement of esoteric Buddhism.

Here, as with the other sundials described in this chapter, we are struck by the expertise of Korean horologists and craftsmen in extending the possibilities of solar timekeeping; despite the excellence and complexity of their clepsydral technology, the 'Men of the East' relied above all on sundials for their standard of time.

13. Plummet Sundial (*hyŏnju ilgu*; *hsüan-chu jih-kuei* 縣珠日晷) (Fig. 2.18)

Kim Ton's memoir on the Simplified-Instrument Platform continues,

Sejong sillok, 77: 10a (1437):

They also made [several] Plummet Sundials, as specified further below, each supported on a square stand 6.3 (Chou) inches on each side, with a column set up in its northern part. On the southern part of the stand there was a pool (*ch'ih* 池); north of the column was engraved a cross, with a hanging weight suspended, from the column, above it. In this way it was unnecessary to use water for levelling; (the instrument) adjusted itself automatically. The Hundred Intervals (*k'o*) were drawn round a small disc with a diameter of 3.2 (Chou) inches; this had a handle jutting out in a slanting position from the column. There was a hole in the centre of the disc threaded through with a thin thread, attached to the top of the column above, and to the southern part of the stand below. The time was given in double-hours and intervals by the location of the shadow of the thread.

Fig. 2.18 is a reconstruction drawing of this instrument, done on the basis of Kim Ton's description. In the stand is a pool, perhaps used (as might have been also the case for several of the other instruments described above) to float a compass-needle for orientation. The main column of the sundial is placed near the north end of the stand's centre-line. From its top a bracket, projecting towards the north, holds a plumb-line and bob suspended over an incised cross to ensure that the instrument is properly levelled. A disc, inscribed on both faces with the double-hours and intervals, is fixed in the equatorial plane (i.e. at an angle of about $37\frac{1}{2}°$ to the vertical) on a short handle projecting out from the main column.

Fig. 2.17. Scaphe sundial: Korean, seventeenth-century.
The tip of the gnomon-shadow indicates time of day by its position in relation to the engraved hour-circles (which are coplanar with the gnomon-axis) and seasonal time in Fortnightly Periods (*ch'i*) by its position in relation to the fortnightly solar-declination parallels (the planes of which are perpendicular to that axis). The circles and parallels are identified by inscriptions on the flattened rim of the scaphe.

We assume that the double-hours and intervals were arranged as 96 *k'o* with 'remainder *fen*' distributed amongst the twenty-four half double-hours, as on the Hundred-Interval ring, no. 3 above. A thread runs from the top of the column, at right angles to the face of the disc and through a hole in its centre, to an attachment point (a tension-screw?) at the southern edge of the base, thus forming a pole-pointing string-gnomon.[85]

Jeon treats this instrument and the Horizontal Sundial, no. 15 below, together,

[85] We discuss the origins and implications of the polar thread-gnomon in connection with the Horizontal Sundial, no. 15 below; see pp. 86–8.

and tends to conflate the two;[86] but it is evident from Kim Ton's descriptions that the two instruments are quite different in conception.

We have already discussed a later form of the Plummet Sundial in connection with the Sun-and-Stars Time-Determining Instrument; see Fig. 2.13 and p. 60 above.

14. Travel Clepsydra (*haengnu*; *hsing-lou* 行漏)

Kim Ton writes,

Sejong sillok, 77: 10ab (1437):

On cloudy and overcast days it was difficult to know the time. So they made Travel Clepsydras (*haengnu*), small and simple things having one water-receiving vessel and one outflow vessel, and a siphon [or outflow tube] (*k'o-wu* 渴烏, 'thirsty crow') to make the water drip. They could be used between the *tzu* 子 and *wu* 午 double-hours, or between *mao* 卯 and *yu* 酉 [i.e. they would run for only six double-hours at a time].

Of the Small Sun-and-Stars Time-Determining Instruments, the Plummet Sundials, and the Travel Clepsydras, there were several of each. They were distributed among the army corps, and those that were left over were kept at the Observatory.

This brief description admits of two quite different interpretations, and again we find ourselves wishing that Kim Ton had given us more information.

On the one hand, this instrument might have been a simple and therefore 'portable' (*hsing* 行) inflow-float clepsydra having one outflow tank (perhaps of leather, for even greater portability), a siphon (or perhaps a simple outflow tube),[87] and an inflow vessel with a float and indicator-rod. This would correspond to a type of clepsydra known as 'Mr Ch'en's *Chou-li* style ancient clepsydra'.[88]

[86] Jeon, STK, p. 49.

[87] The statement has often been repeated (e.g. in Jeon, STK, p. 148; and, more cautiously, D. R. Hill, *Arabic Water-Clocks*, p. 5) that a siphon used with a clepsydra outflow vessel of the non-constant-level type can partly compensate for the effects of unequal and declining water pressure. This, if true, would make the use of a siphon advantageous as compared with the use of a simple outflow tube located at the bottom of the tank.

In a series of experiments kindly conducted for us by Sir Brian Pippard of the Cavendish Laboratory, Cambridge University, this alleged effect did not manifest itself. The outflow rates for both siphons and simple outflow tubes were found to be nearly the same (within 1 per cent) under a variety of conditions. A siphon can compensate for unequal and declining pressure in the outflow vessel only if the siphon itself is carried on a float within the vessel. We know of no East Asian clepsydra that included any such arrangement.

[88] Needham, SCC III: Fig. 138, top right, and Fig. 144, right. Jeon, STK, p. 64, favours this interpretation, basing his description on SCC III: Fig. 144.

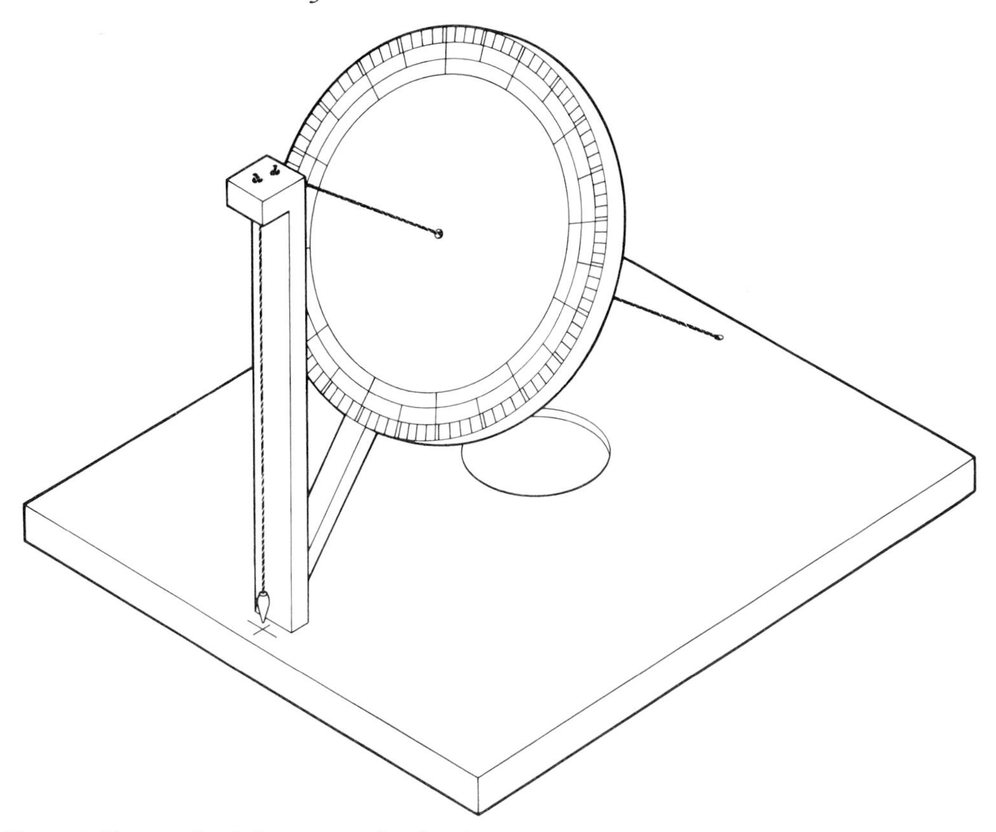

Fig. 2.18. Plummet Sundial: reconstruction drawing.

The equatorial dial-plate (diameter 3.2 Chou inches; 78.5 mm) is graduated with twelve double-hours (*shih*), and with twenty-four half-double-hours each comprising $4\frac{1}{6}$ intervals (*k'o*, i.e. $\frac{1}{100}$ of a day-and-night).

An axial thread at the Seoul polar elevation of 37.5° forms the gnomon, and a plumb-line is provided for levelling. In the 6.3-Chou-inch (155 mm) square base-plate there is a pool which may have been used for floating a compass-needle.

On the other hand, *hsing-lou* (or *hsing k'o lou* 行刻漏, 'travelling "interval" clepsydra') has been considered another name for a steelyard clepsydra, *ch'eng-lou* or *ch'eng k'o lou* 稱刻漏, 'balance "interval" clepsydra'.[89] In this usage the word

[89] Needham, SCC III: 317–18 and Fig. 138 (also Needham, Wang, and Price, HC, 88–94 and Fig. 38), Types C and D. See also John H. Combridge, 'Chinese Steelyard Clepsydras', *Antiquarian Horology*, Spring 1981, *12.5*: 530–5 (which, however, is concerned with principles of operation and does not go into questions of portability); Richard P. Lorch, 'Al-Khāzinī's balance-clock and the Chinese steelyard clepsydra', *Archives Internationales d'Histoire des Sciences*, June 1981, *31*: 183–9, for knowledge of which we are indebted to Anthony J. Turner, *The Time Museum Catalogue, Volume I, Part 3: Water-clocks, Sand-glasses, Fire-clocks* (Rockford, Illinois, The Time Museum, 1984), p. 20, n. 119; and André W. Sleeswyk, 'The Celestial River: A Reconstruction', *Technology and Culture*, 1978, *19.3*: 423–49. Combridge and Lorch differ markedly from Sleeswyk in their interpretations of this instrument. One of us (JSM) is now preparing a critique of the differing approaches and a re-examination of the entire problem, to be published under the title 'The Chinese Steelyard Clepsydra Reconsidered'.

hsing might refer to the (manual) movement of the weight along the beam, as distinct from the 'travel' of the user and/or the 'portability' of the instrument (or perhaps the word is intended to convey both meanings). There was also a *ma-shang* 馬上 ('horseback' or 'rapid') *hsing-lou*, which may have been a steelyard clepsydra especially designed either for extra portability, or for use when the user was in a hurry to know the time (cf. the next item below), or for timing short intervals; but we will not now investigate that doubtful question further. Nothing in Kim Ton's description excludes the possibility that the *haengnu* was a steelyard clepsydra, but nothing specifically suggests it, either. Therefore, until such time as additional information may come to light, we find no basis for choosing between the two possibilities.

15. Horizontal Sundial (*ch'ŏnp'yŏng ilgu*; *t'ien-p'ing jih-kuei* 天平日晷) (Fig. 2.19)

Following his digression on the Travel Clepsydra, Kim Ton returns to the subject of sundials. He writes,

Sejong sillok, 77: 10b (1437):

> (For occasions when) it was necessary to tell the time quickly (*ma-shang pu-k'o pu-chih shih* 馬上不可不知時), (a number of portable) Horizontal Sundials were made. These were on the whole similar to the Plummet Sundials, but they had only a pool chiselled out in the southern part of the base and a column set up to the north. A thread ran from the centre of the base to the top of the column. They were held up and pointed to the south; this was the only difference.

This description also contains a certain amount of ambiguity. On the one hand, Kim says that this sundial was 'on the whole similar to the Plummet Sundial', which would lead one to expect an equatorially mounted dial. On the other hand, the description specifies only a base-plate and pool, a column, and a polar thread-gnomon. The instrument's name, '*t'ien-p'ing*' (sky-level), seems to indicate that it was planar. The problem lies in what Kim Ton may have meant by 'on the whole similar'.

If this instrument was what it seems to have been, a portable horizontal-plane sundial (Fig. 2.19), then it might have been the ancestor of the thread-gnomon 'pocket-watch' sundials that became popular throughout the world in the seven-

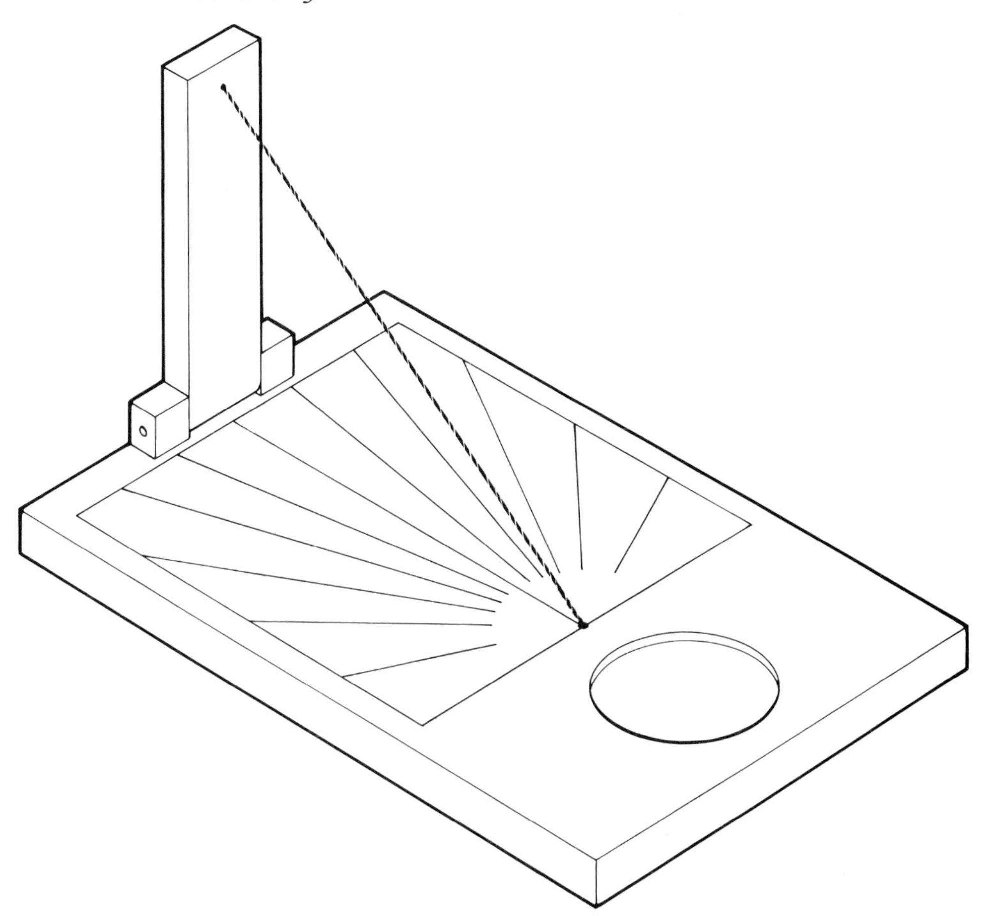

Fig. 2.19. Horizontal Sundial: reconstruction drawing.

The horizontal dial-plate has a pool for levelling, and perhaps for floating a compass-needle. The plate is shown as graduated for half-double-hours. A thread at the Seoul polar elevation of 37.5° forms the gnomon, its upper end being supported by an upright which has conjecturally been shown as hinged for portability.

teenth century.[90] This early model did not necessarily fold up, however.[91] The 'pool' could have served both to float a compass-needle for orientation, and as a water-level of approximate accuracy (even if it lacked the usual side-channels of the water-levels of larger instrument bases); it could easily have been filled with a small amount of water for use, and emptied thereafter, so that the instrument could if necessary have been used even literally 'on horseback' (though not, of

[90] Needham's Type A; SCC III: 310ff and Fig. 133.

[91] Jeon, STK, p. 49, says that it did fold up, but the text is not specific on this point.

course, on a *moving* horse). If not intended for levelling, the 'pool' might conceivably have been a miniature scaphe with a pin-gnomon; our assumption that this instrument had a thread-gnomon is based on Kim Ton's statement that it was 'similar' to the Plummet Sundial. If, as we assume, there was no equatorial dial or scaphe, time-scale lines for the double-hours and their halves could easily have been inscribed empirically on the base-plate itself by simple reference to time-readings on an existing equatorial or scaphe sundial. The assumption that the thread-gnomon horizontal-plane sundial was a Jesuit introduction to the Chinese culture-area may now be dismissed.[92]

The use of thread-gnomons having been established in the Simplified Instrument and its derivatives (see pp. 44–70 above), the replacement of the both-directions pole-pointing gnomon of the ordinary Chinese equatorial sundial[93] by a polar thread-gnomon as in the early equatorial Plummet Sundial (no. 13, Fig. 2.18 above) would have been a natural development. From there it was but a short step to the portable horizontal-plane model described in this section (which, like all horizontal-plane sundials, had the advantage that its face was kept in full sunlight all year round).

Yang Yü 楊瑀, in his 'Conversations in the Mountain Retreat Concerning Recent Events', written in Hangchou *c.* 1360, describes a Simple Small Sundial (*chien-i hsiao jih-kuei* 簡易小日晷), scarcely 9 cm long, that might well have been the ancestor of the Horizontal Sundial described here.[94] Yang's account contains the tell-tale phrase '*ma-shang*'; he says that 'on horseback [or 'rapidly'], one could hold it in one's hand to tell the time; it was very convenient when travelling'. Unfortunately, Yang provides no details on whether his small sundial was equatorial or horizontal-planar, nor on what type of gnomon it had.

[92] Needham, SCC III: 310.

[93] *Ibid.* Fig. 131. The both-directions pole-pointing stylus-gnomon of the two-faced equatorial sundial was presumably well known throughout the Chinese culture-area at this time. See *ibid.* 308–9 for a translation and discussion of part of an essay by Tseng Min-hsing 曾敏行 (1176), ascribing the invention of this device to his father (or grandfather?), Tseng Nan-chung 曾南仲, *c.* 1130.

[94] *Shan-chü hsin-hua* 山居新話, 16b. The passage has been translated by Herbert Franke in 'Beiträge z. Kulturgeschichte Chinas unter der Mongolenherrschaft', *Abhandlungen für die Kunde des Morgenlandes*, 1956, *32*: 1–160, p. 63. See also Needham, SCC III: 311.

16. South-Fixing Sundial (*chŏngnam ilgu*; *ting-nan jih-kuei* 定南日晷) (Fig. 2.20)

Kim Ton continues,

Sejong sillok, 77: 10b (1437):

If one wished to investigate the heavens in order to know the time, it had not (previously) been possible to avoid using a compass, and therefore human intervention. So an instrument was made called the South-Fixing Sundial. It was (so called because) although a magnetic compass was not used, it still aligned itself on the north–south axis. The stand was 1.25 feet long. At both ends, for a distance of 2 inches, it was 4 inches wide; the central 8.5 inches of the stand was 1 inch wide. At the centre there was a round pool 2.6 inches in diameter, having water-channels reaching to both ends and encircling the columns. There were two columns: the north column 1.1 feet high, the south column 5.9 inches high. On the north column, 1.1 inches below the top, and on the south column 3.8 inches below the top, each carried a pivot (*chu* 軸) to receive a Component-of-the-Four-Displacements ring [mobile declination ring], which could rotate from east to west. Half of it was graduated with the degrees of the celestial circumference, each degree having four *fen*. From North 16^{d} to 167^{d} there was a slot in the ring (so that it was) like the double rings (in armillary spheres). The rest was a solid ring. On the inside of the ring was engraved a median line. In its lower part was a square aperture. Transversely there was fixed a cross-strut with a slot 6.7 inches long in the middle, carrying a sighting-alidade (*kuei-heng* 窺衡). At the top it passed between the split rings, and at the bottom it reached nearly to the solid part of the ring. It could decline and rise to north and south.

There was set up a (fixed) horizon ring, level with the top of the south column, used for equalising (the instrumental azimuths of) sunrise and sunset at the summer solstice. There was also set transversely, below the level of the horizon (ring), an (equatorial) half-ring, the inside surface of which was marked with 'intervals' (*k'o*) opposite the square aperture (in the lower part of the declination ring).

At the northern end of the stand was inscribed a cross, over which a plummet was hung from (a projecting portion of) the northern pivot axle; this was for the purpose of levelling.

Every day the sighting-alidade was reset according to the sun's north-polar distance. The rays of the sun form a round spot (on the median line of the declination ring, when this ring is correctly positioned).

Then, looking down through the square aperture, the time can be seen on the intervals of the (equatorial) half-ring. (The instrument) thus automatically fixes the south and tells the time.

Fifteen of these were made, ten being executed in bronze. After some years they were finished.

The construction and use of this instrument are explained in the caption to Fig. 2.20 and need no additional comment here. Our understanding of the South-Fixing Sundial differs in important respects from that of Jeon.[95]

17. The Chou Foot-Rule (*chuch'ŏk*; Chou ch'ih 周尺)

After discussing the South-Fixing Sundial, Kim Ton concludes his memoir on the Simplified-Instrument Platform with the long postface that we have already quoted at the beginning of this chapter. Following Kim Ton's memoir, the authors of the Veritable Records add a long paragraph (77: 11ab) discussing the foot-rule used as the standard measure in the making of all the instruments described above.[96] It specifies that the measure known as the Chou Foot-Rule was to be used for all astronomical instruments. Knowledge of this measure was preserved in the writings of Chu Hsi 朱熹 and Ssu-ma Kuang 司馬光, and moreover old foot-rules from the [Later?] Chou period had been preserved. These were used in scaling King Sejong's instruments.

Jeon gives a derived figure of 21.27 cm for the length of the Chou foot of King Sejong's time.[97] The question of the *chuch'ok* would seem to require further investigation, however, in the light of a recent Chinese study.[98] I Shih-t'ung 伊世同 has shown that a bronze gnomon formerly at Peking and now at the Purple Mountain Observatory in Nanking was scaled to the 'celestial foot' (*liang-t'ien ch'ih* 量天尺) of the Ming Cheng-t'ung 正統 reign-period (1436–49). That measure, 24.525 cm in length, was ultimately derived from the 'iron foot' (*t'ieh ch'ih* 鐵尺) of the Later Chou dynasty (557–80). As the Ming bronze gnomon dates from exactly the same time-period as the re-equipping of King Sejong's observatory, it seems not unlikely that the same measuring-scale was used in both cases.

All Chinese foot-rules were divided decimally into *ts'un* (寸). The usual translation of *ts'un* as 'inches' may in some contexts tend to mislead by its implication of 'twelfths of a foot'. But in the present context it may actually be helpful

[95] Compare Jeon, 'Yissi Chosŏn ŭi sigye chejak sogo', pp. 57–9.

[96] Partly tr., and discussed, in Jeon, STK, pp. 131–4.

[97] *Ibid.* p. 134.

[98] I Shih-t'ung 伊世同, 'Liang-t'ien-ch'ih k'ao' 量天尺考 (On the 'foot' used in celestial measurements), *Wen-wu* 文物, 1978.2 10–17; summarised by Xi Zezong in 'Chinese Studies in the History of Astronomy, 1949–1979', p. 467.

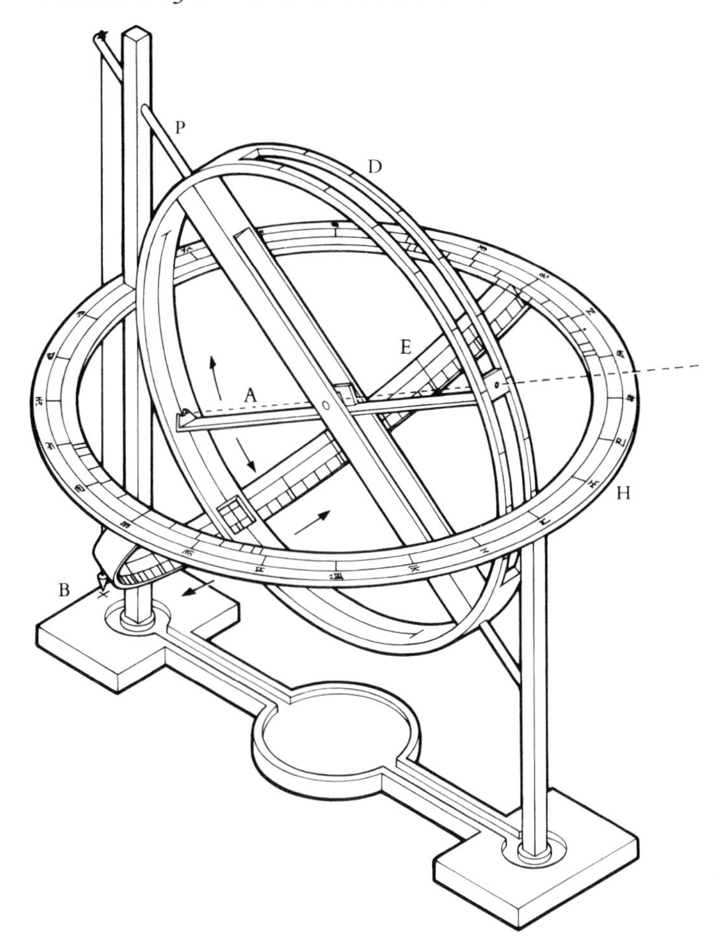

Fig. 2.20. South-Fixing Sundial: reconstruction drawing.

The dimensions of the base, pillars, and axial strut of this sundial are recorded in sufficient detail to allow preparation of a full-scale drawing, from which this perspective view was derived.

The mobile solar-declination ring D (outside diameter 9 Chou inches; 221 mm) is graduated for the sun's polar distance. The fixed horizon ring H (outside diameter 11.6 Chou inches; 284 mm) is graduated to indicate the twenty-four Chinese azimuth directions as well as the azimuths of sunrise and sunset during the twenty-four Fortnightly Periods. The fixed equatorial half-ring E (inside diameter 9.4 Chou inches; 230 mm) is graduated to show twelve daytime half-double-hours, paired as full double-hours, and subdivided to show $4\frac{1}{6}$ intervals (*k'o*) in each half-double-hour. A plumb-bob B, for levelling, is not quite hidden from view by the northern part of the equatorial half-ring.

When in use, the centrally pivoted alidade A was set to the sun's polar distance for the current Fortnightly Period, and the base of the instrument was set in azimuth so that the observed points of sunrise and sunset fell symmetrically on either side of the instrument's north–south centre-line, thus fixing its true direction without use of a compass.

For time-determination, the declination ring was then rotated on its polar axis P to the position where the spot of sunlight from the small hole at the south end of the alidade fell along the centre-line of the alidade onto the median line of the declination ring or, at dates near the equinoxes, through the square aperture in this ring onto the time-scale of the equatorial half-ring. Solar time was then read on the equatorial half-ring, at the centre of the square aperture.

towards visualisation of the dimensions of the instruments, since a 'Chou inch' of 2.4525 cm does not differ greatly from the modern Western inch of 2.54 cm.

18. Rain-gauges (*ch'ŭgu ki*; *ts'e-yü ch'i* 測雨器) (Fig. 2.21)

In *Sejong sillok*, 96: 7ab, and also in the account of the re-equipping of King Sejong's Royal Observatory in *Chŭngbo munhŏn pigo*, 2: 32a, it is recorded that rain-gauges were made in 1442. In the Chinese tradition, no clear distinction was made between astronomical and meteorological phenomena; both were included in the category of *t'ien-wen* 天文. Thus rain-gauges also came under the authority of the Sŏun Kwan. Just as astronomical data was employed, by means of astrology, in the conduct of state affairs, so also was rainfall data put to practical use: it was intended to improve the assessment of agricultural land taxes.

King Sejong's rain-gauges have been described as the first instruments ever made for the scientific measurement of rainfall.[99] In fact, however, it appears that they were derived from earlier Chinese instruments. There is clear evidence that rain-gauges were known in China during the thirteenth century,[100] and Chu K'o-chen has found that they may have been in use in China as early as the T'ang period.[101] King Sejong's rain-gauges have been described fully by Jeon[102] and need no further comment here; we mention them, and include an illustration (Fig. 2.21) of an exact replica of one of them made during the eighteenth century, for the sake of completing our account of the instruments of the Korean Royal Observatory in King Sejong's time.

Looking back over the instruments described in these pages, we see that Kim Ton's boast, 'even Kuo Shou-ching of the Yüan could have done no better', was no idle one. In the following chapter we shall see how these remarkable instruments were variously lost, preserved, repaired, or replaced over the course of the next three centuries of the Yi period.

[99] Jeon, STK, p. 108.

[100] Needham, SCC III: 471–72.

[101] Chu K'o-chen 竺可楨, 'Lun ch'i-yü chin-t'u yü han-tsai' 論祈雨禁屠與旱災 (A discussion of praying for rain (by) prohibiting the slaughter of animals, in connection with drought-disasters), originally published in *Tung-fang tsa-chih* 東方雜誌 (The Eastern Miscellary), 1926, *23*.13: 15–18; repr. in *Chu K'o-chen wen-chi* 竺可楨文集 (Collected works of Chu K'o-chen) (Peking: Science Press, 1979), pp. 90–9; see pp. 92–3.

[102] Jeon, STK, pp. 108–11.

Fig. 2.21. Eighteenth-century copy of one of King Sejong's fifteenth-century rain-gauges.

3

FROM MUNJONG TO YŎNGJO (1450–1776)

In this chapter we present a survey of activities concerning the astronomical instruments of the Yi Royal Observatory from the end of the reign of King Sejong through the reign of King Yŏngjo, as a prelude to the two more detailed and specific studies to be found in Chapters 4 and 5.

BACKGROUND

At the death of King Sejong in 1450, the Korean Royal Observatory possessed one of the finest and most complete sets of astronomical instruments in the world. For the next 150 years efforts were made to keep those instruments in good repair, and from time to time new instruments were constructed to replace or augment the older ones. The Bureau of Astronomy faithfully carried out its calendrical, meteorological, and observational functions, producing a record of celestial events comparable to that contained in the annals of the Ming Dynasty.[1] At the end of the sixteenth century and into the early seventeenth century, events compelled an almost total renewal of the Royal Observatory's equipment. Yet until the reign of King Hyojong (1649–59), the Korean tradition of astronomy and astronomical instrument-making produced little that was innovative. For 200 years after King Sejong's death, then, his great legacy of instrument-making was upheld but not substantially augmented.

There is nothing very surprising in this. In China, too, the thirteenth-century instruments of Kuo Shou-ching, with a few fifteenth-century copies and additions, served as the main equipment of the Imperial Observatory until they were superseded by the new-style instruments made by members of the Jesuit China Mission in the seventeenth century. In both China and Korea, courts

[1] See, for example, the article by Hsi Tse-tsung (Xi Zezong) and Po Shu-jen (Bo Shuren) cited in our Introduction, n. 7.

possessing instruments comparable to the best in the world had little incentive to improve upon them. In Korea, moreover, instruments received as a legacy from so venerated an ancestor as King Sejong were to be revered, not replaced. In addition, from the time of King Munjong, Sejong's successor, until the end of the sixteenth century, Korea received relatively little external stimulus for change in this field. The excellent relations with the Ming court that had been established by 1400 by King T'aejo and his immediate successors continued. But the Ming period, as we have just noted, was not an era of great innovation in astronomy, so that the embassies exchanged by the two courts transmitted relatively little information on new types of instruments.

The re-equipping of the Royal Observatory in the fifteenth century was undertaken largely in response to the consequences of the fall of the Yüan Dynasty and the Kingdom of Koryŏ. The newly established Kingdom of Chosŏn felt a need to provide itself with a suitable set of up-to-date astronomical instruments. The second great period of astronomical invention during the Yi period also took place against a background of extraordinary and highly disruptive events: the successive invasions under the Japanese tyrant Hideyoshi and the newly confederated Manchu tribes, the fall of the Ming Dynasty, and the impact of new ideas and techniques imported from the West via the Jesuit missionaries in China.

The Yi court found the threat of Japanese invasion hard to take seriously at the end of the sixteenth century. Almost no preparations for national defence were undertaken before Hideyoshi's troops, equipped with very effective firearms, arrived on Korean shores in 1592. Even had a decision been made to mobilise for defence, it is unlikely that Korea's administrative and logistical structure would have provided an adequate foundation for a successful defence effort.[2] When the invasions began, the hasty evacuation of Seoul by the Yi court gave rise to rioting and looting by an urban populace that seems to have felt betrayed by its rulers; it is likely that many of King Sejong's instruments were destroyed in the burning of palace buildings even before the Japanese occupied Seoul.[3] Six years of subsequent warfare saw the destruction of almost all that might have remained.[4]

An official calendar, a coterie of Astronomers-Royal, and a set of astronomical instruments were, by a tradition extending back to the legendary age of Yao and

[2] Henthorn, *A History of Korea*, pp. 178–9; this point was also stressed to us by Dr Gari Ledyard (private comm.).

[3] Henthorn, *A History of Korea*, p. 180.

[4] Yoshi S. Kuno, *Japanese Expansion on the Asiatic Continent*, 2 vols. (Berkeley, 1937–40), I: 340–2.

Shun, indispensable to the prestige of an East Asian ruler. Thus it was a matter of priority, after the Japanese had been driven from Korea, to rebuild the Royal Observatory. This was to be done largely on traditional lines; as we shall see below, instrument-making in the early seventeenth century involved little more than an attempt to restore or replicate the instruments of King Sejong's time.

Hardly had the task been taken up, however, when a fresh disaster struck. Sensing the decrepitude of the Ming, the tribes of Manchuria had begun to form a grand confederation by about 1615. The pro-Ming sentiments of the Yi court made Korea a natural focus of Manchu concern. With the overthrow of the Ming already a feasible goal of Manchu expansion, they sought to secure their southern flank by obtaining the submission of Korea. Thus in 1627 began a decade of Manchu attacks on northern Korea, leading to a series of wars that were hardly less destructive than the Hideyoshi invasions, and were doubly dispiriting for coming so close on the heels of the earlier disasters.[5] As was true in other fields as well, the recovery of the Royal Observatory slowed during the decade of the Manchu invasions, though it did not entirely come to a halt. In 1637 the Yi court was placed in the humiliating position of concluding a separate peace with the Manchus while still trying to maintain its loyalty to the Ming.

The fall of the Yüan Dynasty in 1368 had been followed by the fall of Koryŏ in 1392; while the Chosŏn Kingdom did not follow the Ming into extinction after the Manchu victory of 1644, the effect of dynastic change in China on the Korean tributary state was nevertheless profound. After several years of indecision and confusion, in 1651 the Yi throne ritually signified its capitulation to the newly established Ch'ing dynasty by adopting the Ch'ing calendar.[6]

All of Yi theoretical astronomy had been based on the Shou-shih calendar of Kuo Shou-ching (1280) and its subsequent Chinese and Korean redactions, just as King Sejong's instruments were largely based on Kuo Shou-ching's technology.[7] The adoption by the Ch'ing of a radically new calendar therefore demanded a response from the Korean Bureau of Astronomy, and provided an opportunity for a period of creativity analogous to that of King Sejong's time – which, as we have seen, was in large part a response to the establishment of the Ming. The reigns of Kings Hyojong (1649–59) and Hyŏnjong (1659–74) saw a renaissance of Korean astronomy, including the production of a remarkable series of astronomical clocks.

[5] Hatada, *A History of Korea*, pp. 79–81. [6] Jeon, STK, p. 83. [7] *Ibid.* pp. 78–83.

Table 3.1. *Chronological summary of instrument-making and related activities in the middle Yi period, 1450–1777*

1455–69	Striking Clepsydra in the Porugak breaks down in 1455 and is repaired in 1469; it might have been operated manually during the interval between these years
1489	Repairs to the astronomical and meteorological observatories
1494	King Sŏngjong orders the casting of a new Small Simplified Instrument (*so kanŭi* 小簡儀)
1505	Striking Clepsydra is moved to the Ch'angdŏk Palace
1525	Astrolabe (*mongnyun* 目輪 or *kwanch'ŏn ki* 觀天器) made on the basis of designs imported from China
1526	Sejong's water-operated demonstrational armillary sphere repaired
1534–6	Striking Clepsydra repaired once again, and a duplicate made; the duplicate instrument is reinstalled in the Porugak
1549	Sejong's water-operated demonstrational armillary sphere replaced by a copy of the original instrument; the new instrument is installed in the Hongmun'gwan
1554	The Hŭmgyŏnggak is rebuilt on its old foundations, and a copy of Sejong's Jade Clepsydra is installed. (The original Hŭmgyŏnggak and Jade Clepsydra had been lost in a fire which destroyed three buildings of the Kyŏngbok Palace in 1553)
1592	Most of the instruments at the Royal Observatory and elsewhere in the palace are destroyed in the course of the Hideyoshi invasion. The Simplified-Instrument Platform, the Sun-and-Stars Time-Determining Instrument, and the Striking Clepsydra in the Porugak (possibly damaged) survive
1601	Yi Hangbok is ordered to make a new armillary sphere and celestial globe (replacing the lost 1549 copy of Sejong's old instrument)
c. 1601	Yi Kyŏngch'ang writes essay describing operation of an armillary sphere (presumably automatically rotated)
1607	Striking clock (perhaps of Western design) presented by Korea to Japan
1614	The Hŭmgyŏnggak is rebuilt by Yi Ch'ung, in a new location; there is, however, no indication that the Jade Clepsydra that had been housed in the old Hŭmgyŏnggak is replaced
1614	Yi Ch'ung also rebuilds the Porugak in a new location, the Striking Clepsydra (1536 copy) is restored, at least to some extent using parts that had survived the invasions
1631	Telescope imported from China by Chŏng Tuwŏn. There is, however, no indication that it was ever used for astronomical purposes. Chŏng also brought back with him a Western clock
1636	'New Model Horizontal Sundial' (*sinpŏp chip'yŏng ilgu* 新法地平日晷), based on an instrument of J. A. Schall von Bell, imported from China
1651	Ch'ing Shih-hsien calendar (時憲曆) adopted by the Yi court
1653	Jackwork of the Striking Clepsydra dismantled and the instrument converted to manual operation and time-annunciation
1650s	Yu Hŭngbal studies weight-driven clock imported from Japan
1657	Hong Ch'ŏyun makes a *sŏn'gi okhyŏng* (璿璣玉衡), (water-operated) armillary sphere. It does not work successfully
1657	Ch'oe Yuji makes a new *sŏn'gi okhyŏng*, installed in the Clepsydra Bureau. The new instrument works well; two or more copies may have been made
1664	Song Iyŏng and Yi Minch'ŏl ordered to renovate Ch'oe Yuji's automatically rotated armillary sphere(s)
1669	Song Chun'gil submits a petition asking that King Sejong's old Jade Clepsydra be rebuilt. In response, Song Iyŏng and Yi Minch'ŏl make one or more new (or renovated) instruments. At the end of this decade of activity, at least two astronomical clocks emerged: a weight-driven one associated with Song Iyŏng, and a water-operated one associated with Yi Minch'ŏl
1669	Kim Sŏkchu writes a memoir describing the clocks of Song and Yi

1660s–70s	King Sejong's scaphe sundials, lost in the Hideyoshi invasions, are replaced. These were palace instruments; the old scaphe sundials for public use apparently were not replaced
1687	Song Iyŏng's weight-driven astronomical clock repaired by Yi Chinjŏng; Yi Minch'ŏl repairs his own water-operated astronomical clock
1704	An Chungt'ae makes a copy (larger) of Yi Minch'ŏl's armillary clepsydra; installed in Kyujŏng Pavilion
1715	Copy of Chinese Western-style weight-driven clock made for the Bureau of Astronomy
1723	Another Western-style weight-driven clock imported from China
1732	An Chungt'ae repairs his 1704 copy of Yi Minch'ŏl's armillary clepsydra. King Yŏngjo writes a memoir describing it
1759–61	Hong Taeyong makes instruments, including a weight-driven armillary sphere, a weight-driven striking clock, and a water-operated celestial globe, for his private observatory
1777	Yi Minch'ŏl's original armillary clepsydra repaired once again

Against this background, we shall here briefly review the works of the Yi instrument-makers before the reign of King Hyojong. We shall then look in more detail at the innovative instruments that began to be produced during his reign, and at their subsequent history.

Table 3.1 provides a chronological summary of instrument-making and related activities during the period under consideration in this chapter.

EVENTS PRIOR TO KING HYOJONG'S ACCESSION

1. The Striking Clepsydra

King Sejong's Striking Clepsydra (see Chapter 2, no. 1, pp. 23–44), housed in the Porugak (Annunciating Clepsydra Pavilion), broke down in 1455. It is not clear whether the instrument ceased working altogether, or whether only its automatic apparatus failed so that its time-annunciation function had once again to be handled by officials of the Clepsydra Bureau. In any case, the instrument was repaired in 1469; in 1505 it was moved from the Porugak to the Changdŏk Palace. In 1534 it broke down again, and repairs were ordered; at the same time, a duplicate was made and reinstalled in the Porugak. Both the repaired instrument and the duplicate were in operation in 1536. The much-repaired instrument from King Sejong's time was destroyed in 1592 in the Hideyoshi invasions. The Porugak, which housed the 1536 duplicate, was also destroyed at the time, but the duplicate instrument (or parts of it) survived, with unspecified damage. When the

Porugak was rebuilt in a new location by Yi Ch'ung 李冲 in 1614, the 1536 duplicate instrument was once again restored.[8]

The Jesuit-designed Shih-hsien 時憲 calendar of the new Ch'ing Dynasty, adopted by Korea in 1651, discarded the old Chinese system of dividing the day-and-night into twelve double-hours and also into 100 intervals (*k'o*) (or, in practice, 96 intervals with distributed 'remainder fractions' (*fen*)), replacing it by a twenty-four-hour system of fifteen-minute quarter-hours in the Western fashion. It also discarded the variable night-watches. As the automatic apparatus of the Striking Clepsydra operated according to the old systems of double-hours with intervals and fractions, and variable night-watches, it was rendered obsolete by the new calendar, and was dismantled in 1653. The clepsydra itself remained in use, but the time was read by officials of the Clepsydra Bureau and announced manually by the sounding of bells, drums, and gongs.[9] Parts of the 1536 duplicate clepsydra, as modified in 1653, survived until the 1950–3 Korean War, when more of it was destroyed. The surviving components, partially restored, now stand in the grounds of the King Sejong Memorial Hall in Seoul (see Fig. 2.2).

2. The Jade Clepsydra

For all its complexity, the Jade Clepsydra (Chapter 2, no. 11, pp. 76–80) of the 1430s apparently remained in good working order until the Hŭmgyŏnggak (Hall of Respectful Veneration) in which it was housed was burned down in a conflagration which destroyed three buildings of the Kyŏngbok Palace in 1553, in the reign of King Myŏngjong. The Hŭmgyŏnggak was rebuilt on its old foundations in 1554, and an exact copy of the Jade Clepsydra was installed therein.[10] Yi Ch'ung, the rebuilder of the Porugak, also rebuilt the Hŭmgyŏnggak in a new location in 1614, but there is no reason to think that the Jade Clepsydra was again restored at that time, and some reason to think that it was not.[11] As late as 1669, Song Chun'gil 宋浚吉 submitted a petition requesting that the clepsydra anciently housed in the Hŭmgyŏnggak should be rebuilt.[12] This petition led not to the rebuilding of the Jade Clepsydra in its original form, however, but to the

[8] *Ibid.* pp. 59–60. *Chŭngbo munhŏn pigo*, commentary on 3: 1b. (Hereafter CMP.)

[9] Jeon, STK, pp. 60, 68, 87–93.

[10] *Ibid.* p. 62.

[11] CMP, 3: 1b. Our view of this has changed since the writing of Needham, SCC IV.2: 521, which states that the Jade Clepsydra had been restored again in 1614.

[12] CMP, 3: 2a.

clockmaking activities of Song Iyŏng 宋以穎 and Yi Minch'ŏl 李敏哲 that we shall examine in detail below; the Jade Clepsydra itself was heard of no more.

3. King Sejong's water-operated demonstrational armillary sphere and related instruments

King Sejong's demonstrational armillary sphere, and presumably the celestial globe associated with it (Chapter 2, nos. 9 and 10, pp. 74–6; as we noted there, one may assume that both were rotated by a common clepsydral mechanism), continued to operate until 1526, when repairs were ordered.[13] By 1549 they apparently had deteriorated beyond repair. In that year the pair of instruments was replaced by replicas,[14] installed in the Hongmun'gwan 弘文館 (College for the Propagation of Literature).[15] The replicas were destroyed during the Hideyoshi invasions.

In 1601, Yi Hangbok 李恒福 was ordered to make a new armillary sphere and celestial globe. These would appear to have been replacements for the lost instruments just mentioned, but the records do not say whether or not the new instruments were equipped with a water-operated drive mechanism.[16] At about the same time, Yi Kyŏngch'ang 李慶昌 wrote a treatise on his *chuch'ŏn* (*chou-t'ien* 周天, 'Revolving Heaven') cosmological system, a refinement of the *hun-t'ien* theory, in which he described a water-operated armillary sphere, referred to as a *sŏn'gi okhyŏng*.[17] We have no further information about these early-seventeenth-

[13] Jeon, STK, p. 67.

[14] *Ibid.* Hulbert, *History of Korea* (I: 334) and Rufus 'Astronomy in Korea' (p. 33) both refer to an armillary clock (in Hulbert, a 'Heaven Measure', in Rufus a Syen-kui-ok-hyeng, i.e. *sŏn'gi okhyŏng*) as having been made in 1549/50; this must refer to the replication of Sejong's armillary sphere. For the term *sŏn'gi okhyŏng* see ch. 2, n. 11; see also below, n. 17.

[15] The Hongmun'gwan, modelled on a T'ang academy of the same name (Hung-wen Kuan) was founded in Korea during the reign of King Sejo (1456–68) and, with various changes in name, continued to function throughout the Yi period. It was responsible for the research, compilation, and publication of a number of important historical encyclopaedic works. Membership in the Hongmun'gwan was by royal appointment, and the College served as an instrument of royal encouragement of literary and scientific projects. During the latter part of the Yi period, its functions became primarily censorial (Dr Gari Ledyard, private comm.).

[16] CMP, 3: 1a.

[17] Jeon, STK, p. 16; Yi Kyŏngch'ang, *Chuch'ŏn tosol* (*chou-t'ien t'u-shuo* 周天圖說, Illustrated explanation of the Revolving Heaven theory). By the middle of the seventeenth century the term *sŏn'gi okhyŏng* usually denoted an armillary sphere rotated automatically by a clepsydral apparatus that might also include jackwork and other time-annunciating mechanisms; this sense of the term applies also in the case of Yi Kyŏngch'ang's early-seventeenth-century instrument.

century instruments, which would have been superseded by the armillary instruments of the 1650s and 1660s.

4. Miscellaneous instruments

In 1494, King Sŏngjong ordered the construction of a new Small Simplified Instrument[18] (see Chapter 2, no. 7, pp. 68–70). This small version of Kuo Shou-ching's equatorial torquetum apparently supplemented, rather than replaced, those made during the re-equipping of the Royal Observatory during King Sejong's time. King Sŏngjong's instrument survived the Hideyoshi invasions and was installed in the Porugak in 1614;[19] its subsequent fate is unknown.

An astrolabe (*mongnyun*; *mu-lun* 目輪) was made by Yi Sun 李純 in 1525, on the basis of diagrams, presumably of Arabic derivation, imported from China.[20] This is the first astrolabe of which we have been able to find any record in Korea. Jeon describes it as having been double-sided, and illustrates such an astrolabe, said to be the actual instrument made by Yi Sun.[21] A Chinese double-sided astrolabe, dated 1681, is illustrated in the *Huang-ch'ao li-ch'i t'u-shih* 皇朝禮器圖史 (Illustrated History of the Ritual Implements of the Imperial Court).[22] Ordinary astrolabes, evidently of Korean origin but undated, are illustrated by Rufus.[23] Thus we see that astrolabes were in use in Korea during the Yi period, but one may assume that in Korea, as in China, they did not achieve the prominence in astronomical observation that they had in the Islamic world and in Europe.[24]

Sundials, the most characteristic of Korean timekeeping instruments, continued to be produced throughout the Yi period; but most of the basic designs had already been established in King Sejong's time.[25] 'Pocket' models, presumably based on the early-fifteenth-century Horizontal Sundial (Chapter 2, no. 15,

[18] CMP, 2: 33a.

[19] CMP, 3: 1b, commentary.

[20] CMP, 2: 33a; Jeon, STK, p. 73; Rufus, 'Astronomy in Korea', pp. 32–3; Rufus and Lee, 'Marking Time in Korea', p. 255. Rufus refers to this instrument as a *kuan-t'ien ch'i* 觀天器. That term could perhaps be an alternative name for the astrolabe; as of 1267, the Chinese name for the astrolabe, newly imported from the Islamic world, had not been decided upon. See Needham, SCC III: 272–4. The phrase *kuan-t'ien chih ch'i* 觀天之器 appears in Kim Ton's memoir on the Simplified-Instrument Platform of 1437 (Chapter 2, pp. 64–5 above), but Kim clearly used it to mean 'observational instruments' in general; so the apparent similarity of the terms need not be taken to mean that an astrolabe was included among King Sejong's instruments.

[21] Jeon, STK, p. 73; this astrolabe is illustrated in the Japanese edition of STK, i.e. *Kankoku kagaku gijutsu shi* 韓国科学技术史 (Tokyo, 1978), Fig. 1–33.

[22] 3: 15a–16a.

[23] 'Astronomy in Korea', Figs. 31 and 32.

[24] Needham, SCC III: 375–6.

[25] Jeon, STK, p. 46; see also Ch. 2, Table 2.1, nos. 2, 4, 12, 13, 15, and 16.

pp. 86–8), became common from the seventeenth century onwards. A 'New Type Horizontal Sundial' (*sinpŏp chip'yŏng ilgu*; *hsin-fa ti-p'ing jih-kuei* 新法地平日晷), made in China in the late Ming period by Li T'ien-ching 李天徑 after an instrument of J. A. Schall von Bell, was imported into Korea in 1636. Jeon describes this as 'a version of the scaphe sundial on a plate model';[26] i.e. it presumably had not a string-gnomon but a point-gnomon, and a stereographic projection of the scaphe sundial's reticular-grid scale.[27]

The only other notable development in the field of sundial-making during the period under consideration here, to the best of our knowledge, is that the scaphe sundials of King Sejong's time, lost during the Hideyoshi invasions, were replaced in the 1660s and 1670s by order of Kings Hyŏnjong and Sukchong. The old instruments described by Kim Ton in his Memoir on the Simplified-Instrument Platform (Chapter 2, no. 12, pp. 80–2) were simple and sturdy, and scaled with pictures of the twelve double-hour gods rather than with a reticular scale. They were made for public use, and those particular instruments apparently were not replaced. The new scaphe sundials of the seventeenth century replaced other earlier ones that had been made for palace use. They were ornately decorated, and were scaled in the usual fashion, except, however, that their reticular grids were graduated in ninety-six Western quarter-hours, reflecting the influence of the Jesuit Shih-hsien calendar,[28] rather than in double-hours with intervals (*k'o*) and remainder *fen*.

Rufus reports that a sighting-tube of some sort (*kyup'yo*; *k'uei-piao* 窺標) was made during the reign of King Sŏngjong in 1491.[29] This is likely to have been an adaptation, for celestial observations, of a terrestrial surveying instrument.[30] It was not, of course, a telescope. A telescope was imported from China in 1631, but it does not seem to have been used systematically for astronomical purposes. Though it is difficult to imagine that no one yielded to the temptation to point it at the heavens (especially since the Jesuit missionaries in China extolled the marvels of seeing the myriads of stars of the Milky Way through the telescope), the new instrument apparently made no significant impact on Korean astronomy.

[26] STK, p. 49. The Japanese edition of STK (*Kankoku kagaku gijutsu shi*), Fig. 1–22, shows a later version of this sundial. Unfortunately, the illustration is very indistinct, so that little detail can be discerned.

[27] Cf. Needham, SCC III: 311–12 and Fig. 137.

[28] Jeon, STK, p. 47.

[29] 'Astronomy in Korea', p. 32; CMP, 2: 33a.

[30] Jeon, STK, pp. 295–6. A supplementary passage in CMP, 2: 33a implies that this was a telescope, invented before the advent of Jesuit instruments, but that is clearly an erroneous *post-facto* interpretation.

Except for some interest in its possible military applications it was regarded simply as a curiosity.[31]

5. The adoption of the Shih-hsien calendar

The Shih-hsien calendar was adopted by the Ch'ing dynasty in 1645, one year after its assumption of power in China. Based on the 1634 *Ch'ung-chen li-shu* 崇禎曆書 (Calendrical Treatise of the Ch'ung-chen Reign-Period) of Jacob Rho, J. A. Schall von Bell, and others, it was thoroughly Western in its methods. Nakayama describes it as 'the crowning achievement of the Jesuit missionaries'.[32]

As a tributary state of the Chinese empire, the Kingdom of Chosŏn had no choice but to accept the Ch'ing as the legitimate successors of the Ming, and to adopt the Ch'ing calendar as a ritual gesture of fealty. While Chosŏn had capitulated to the as yet unconsolidated Ch'ing Dynasty in the Korean–Manchu Peace of 1637, its sympathies and loyalties still lay with the Ming. This fact was well known to the Ch'ing rulers, and led to an anomalous situation. Korea had to accept the Ch'ing calendar, and did so in 1651; yet the Ch'ing, rightly suspicious of the Yi court's loyalty, were most reluctant to supply Korea with the technical information that was necessary to understand the radically new calendrical system. Korean envoys to China thus had to resort to subterfuge to obtain that information, and full details of the Shih-hsien calendar were not known in Korea for nearly half a century.[33]

We have already seen how the adoption of the new calendar led to the dismantling of the jackwork of the Striking Clepsydra (p. 99 above). The urgent need for new instruments that would conform to the new calendar led to a national programme of clepsydra- and clock-making in the 1650s and 1660s. That programme produced, in addition to several armillary striking clepsydras of relatively traditional design, a remarkable armillary clock with a Western-style weight-driven mechanism.

[31] Jeon, STK, p. 77; see also Ch. 5, pp. 168–9 below.

[32] *A History of Japanese Astronomy*, p. 166; see also Needham, SCC III: 449–50.

[33] Jeon, STK, pp. 83–4. Further details on Korean contacts with the Jesuits in China are given in Ch. 5, pp. 175–9 below.

THE ASTRONOMICAL CLOCKS OF KINGS HYOJONG AND HYŎNJONG

While it is obvious from the information preserved in the Veritable Records of Kings Hyojong and Hyŏnjong, and in the *Chŭngbo munhŏn pigo*, that a great deal of horological activity took place during the 1650s and 1660s, it is by no means clear from the texts just how many instruments were produced, or what roles were played by the several instrument-makers whose names have come down to us from that time. At a minimum, three timekeeping instruments were produced; there might have been as many as six or more. We shall first try to clarify the complicated record of who did what, and then comment on the instruments themselves.

In 1657, Hong Ch'ŏyun 洪處尹 was ordered by King Hyojong to make an armillary sphere (*sŏn'gi okhyŏng*).[34] From the context in which it is mentioned, this instrument must have been rotated automatically by a clepsydral apparatus of some kind. Its timekeeping mechanisms were described as being 'very inaccurate', however, and nothing more is heard of it.[35]

Shortly thereafter, Ch'oe Yuji 崔攸之, the magistrate of Kimje 金堤 County, was ordered by the Hongmun'gwan to make a better one. This was clearly described as having been water-operated: 'The (*sŏn'*)*gi* (*ok*)*hyŏng* was automatically turned round by the action of water.'[36] Ch'oe's instrument worked well; 'When compared with the sundial, it did not deviate in the slightest.'[37] It was 'installed in the Clepsydra Bureau'; two or more copies may have been made.[38]

In 1664, King Hyŏnjong noted that 'It has come to Our attention that the

[34] CMP, 3: 1b–2a. See also Ch. 2, n. 11, and n. 17 in this chapter above.

[35] CMP, 3: 1b–2a.

[36] CMP, 3: 2a; Jeon, STK, p. 68.

[37] *Kuei ch'ih-su wu shao ch'a-wei* 晷遲速無少差違. This was the stock phrase used for any mechanical timekeeping device that worked well; it serves as a reminder of Korean expertise in the making of sundials, and reliance on them as the standard of accurate timekeeping. The equivalent Chinese cliché was 'it agreed with (the phenomena of) the heavens like (the two halves of) a tally' (*yu t'ien chieh ho ju fu-ch'i yeh* 與天皆合如符契也); see for example the *Sui shu* 隋書 reference to Chang Heng's armillary sphere, quoted in Needham, Wang, and Price, HC, p. 101.

[38] Jeon, STK, p. 68, says that 'numerous copies were made'. The source of that information is not given, however. It could be an inference from the phrase *chu lou chü* 諸漏局, interpreted as 'the several clepsydra bureaus' (or 'bureau [containing] several clepsydras'). In this text, however, *chu* 諸 consistently is used as a contraction of (*chih*) *chih yu* (置)之於, '(placed) it within'. Thus the phrase must be taken to mean 'it was [or 'they were'] installed in the Clepsydra Bureau(s)' – in other words, there might still have been two or more copies made of Ch'oe's instrument, but the basis for such an assumption is weakened. If Ch'oe's instrument was made in only one example, the appropriate adjustments would have to be made in the following paragraphs regarding the events of 1664 and 1669.

armillary clock(s) of Ch'oe Yuji installed in the Clepsydra Bureau could be improved in some respects.'[39] Accordingly, Song Iyŏng and Yi Minch'ŏl were ordered to make the necessary improvements; the improved instrument(s) was (or were) then installed in the palace. Unfortunately, the record is unclear about exactly what Song and Yi did. The word used to describe their efforts is *kai-tsao* 改造, which might be understood to mean 'repair', 'rebuild', or 'alter'. Later records, as we shall see, consistently refer to two clocks of this period, a weight-driven one associated with Song Iyŏng and a water-operated one associated with Yi Minch'ŏl. It is difficult to determine whether these were successive modifications of Ch'oe Yuji's instruments or new ones built entirely by Song and Yi in 1664 or thereafter.

In 1669, as noted earlier, Song Chun'gil submitted a petition asking that the old Jade Clepsydra formerly installed in the Hŭmgyŏnggak be restored. This apparently was not done, but instead Yi Minch'ŏl was ordered to cast new armillary spheres 'on the basis of Mr Ts'ai's 蔡氏 commentary on the "Shun-tien" [section of the *Shu ching*, the Book of Documents]'.[40] The *Hyŏnjong sillok* says that both Yi Minch'ŏl and Song Iyŏng were ordered to make instruments at this time. Here again, exactly what took place is not made clear. It could be that Yi simply made new armillary spheres to be fitted (perhaps with the assistance of Song Iyŏng) onto the clock(s) of Ch'oe Yuji that Yi and Song had remodelled in 1664, or otherwise onto the (hypothetical) new clocks that the two of them may have made in that year. However, the brief notice in the *Hyŏnjong sillok* that refers to these events makes it seem likely that one of Yi's armillary spheres was used in a new clock with a weight-driven mechanism made by Song Iyŏng in 1669, and that Yi himself made another new water-operated instrument, incorporating another, and larger, specimen of his new armillary spheres, at the same time.[41] Jeon says that Yi and Song each made clocks in 1669, and labels the surviving instrument that we describe at length in Chapter 4 as 'the Seoul armillary clock of Song I-yŏng and Yi Minch'ŏl'.[42]

[39] CMP, 3: 2a; Jeon, STK, p. 68.

[40] CMP, 3: 2a; Yi was a capable hydraulic engineer, as well as a maker of armillary spheres. He invented a waterwheel-powered water-raising device in 1683. See Jeon, STK, p. 158.

[41] *Hyŏnjong sillok*, 17: 35ab. See also Jeon, STK, pp. 69, 163. The terminology of this passage is of interest. Yi and Song are said, respectively, to have made an 'armillary instrument' (*honch'ŏn ŭi*; *hun-t'ien i* 渾天儀) and a 'self-sounding clock' (*chamyŏngjŏng*, *tzu-ming chung* 自鳴鐘). The latter term appears to have been applied exclusively to weight-driven chiming clocks; see Needham, Wang, and Price, HC, p. 142.

[42] Jeon, STK, pp. 68–9 and Fig. 1.17.

Our view also is that by 1669 an armillary sphere of Yi Minch'ŏl and a weight-driven clockwork mechanism made or adapted by Song Iyŏng had in fact been joined to form a single instrument, and it is that which survives to the present day in Seoul. Our reasons for believing this, upon which we shall elaborate below, are as follows. First, Song Iyŏng was unambiguously ordered to make a 'self-sounding [i.e. Western-style] clock' (*chamyŏngjong*; *tzu-ming-chung* 自鳴鐘) in 1669. Second, that he did produce an instrument with a Western-style weight drive is confirmed in an essay written by Kim Sŏkchu 金錫胄 in that same year. Third, the surviving instrument incorporates a clockwork mechanism that shows features characteristic of early Japanese weight-driven clocks. We know that at least one clock mechanism of that sort would potentially have been available for Song to adapt or copy in 1669. Fourth, the armillary sphere of the surviving instrument conforms quite closely to a detailed description of Yi Minch'ŏl's larger water-operated armillary sphere; moreover, it is almost certainly by a different hand than the clockwork itself – its gearing is of a different and more old-fashioned design. That the armillary sphere and the clockwork of the surviving instrument appear to be the work of two separate individuals supports the belief that the instrument is the product of collaboration between Song Iyŏng and Yi Minch'ŏl.

By 1669 there also existed another armillary instrument, water-operated and larger than the surviving clock, associated with the name of Yi Minch'ŏl alone. Whether or not either, or both, of these instruments was ultimately descended from Ch'oe Yuji's instrument(s) of 1657, one may assume that almost all traces of Ch'oe's work would have been remodelled out of existence by 1669. No further new timekeeping instruments are mentioned during the reign of King Hyŏnjong, and those of Song and Yi apparently represent the culmination of a remarkable fifteen-year period of clockmaking.

YI MINCH'ŎL'S WATER-OPERATED ARMILLARY CLOCK

In 1669, Kim Sŏkchu wrote a memoir describing the clocks of Song Iyŏng and Yi Minch'ŏl. Regarding the latter, he noted that 'the method of making the armillary sphere revolve automatically by the action of running water is traditional'.[43] After describing the various rings of the armillary sphere itself (which we shall do in greater detail below, translating from an essay written by King Yŏngjo), he

[43] CMP, 3: 2a–3a.

went on to say,

Chŭngbo munhŏn pigo, 3: 2a–3a (1669):

A water-tank was set up above the wooden cover of the mechanism, and the water poured down from the clepsydra spout gradually into a little vessel inside the casing, and as it became full it made the wheel turn round. Every day water continued unceasingly flowing at constant speed, so that the Component of the Three Arrangers of Time rotated according to its predetermined speed without the slightest mistake. Also at the side there were toothed wheels arranged in a tier with a ball-run and also a mechanism for announcing the time by strokes on a large bell. When compared with the sundial it was all in perfect agreement.

Yi's clock thus does indeed sound traditional, for this description could apply equally well to the driving and time-annunciating mechanisms of the Jade Clepsydra and the Striking Clepsydra of the 1430s. A supplementary note to Kim Sŏkchu's memoir adds the following information,

Chŭngbo munhŏn pigo, 3: 3ab (1669):

Inside a large casing there was set up the water-delivery tube and the ball-run mechanism (*ling-tao chi-kuan* 鈴道機關). South of the casing there was set up the armillary sphere, with its Component of the Six Cardinal Points and its Component of the Three Arrangers of Time, just as in the old designs. The sun and moon each had their rings, and in the centre there was not an alidade, but instead an earth-model of paper with mountains and seas drawn upon it to represent the surface of the earth. The water-delivery tube was associated with a mechanism connected to the north–south polar axis in such a way that its power rotated the rings.

On the west wall of the casing there was a wooden figure in a niche which sounded a bell, and also several jacks with placards for the double-hours. Everything worked perfectly when compared with the sundial.

The water was contained in a vessel at the top of the casing, and ran down into the delivery tube. All the actions of the machinery were effected by the action of water.

The armillary sphere was alongside the housing for the drive and time-annunciating mechanism. The arrangement of Yi's instrument sounds quite similar to that of the extant Song Iyŏng clock, except that the water-operated mechanism of the former must have been considerably larger than Song's compact weight-drive.

Yi Minch'ŏl was ordered in 1687 to make repairs to his clock 'from the time of King Hyŏnjong';[44] no doubt this refers to the 1669 instrument described above.

[44] CMP, 3: 3a; *Sukchong sillok*, 19: 13a; Jeon, STK, p. 71.

In 1704, An Chungt'ae 安重泰 made a copy of Yi Minch'ŏl's instrument, which was installed in the Kyujŏng Pavilion (Kyujŏnggak; K'uei-cheng Ko 揆政閣, Pavilion for Calibrating the (Motions of) (the Seven) Regulators).[45] In 1732 An Chungt'ae was ordered to make repairs to his duplicate instrument.[46] We learn still more about the design of Yi Minch'ŏl's instrument, as copied by An Chungt'ae, from an essay entitled 'Memoir on the Kyujŏng Pavilion' (*Kyujŏnggak ki* 揆政閣記) written by King Yŏngjo himself in 1732, after An had completed the repairs to his duplicate instrument,

Chŭngbo munhŏn pigo, 3: 6b–7a (1732):

The design of the armillary sphere in the Kyujŏng Pavilion (was as follows):

The horizon single ring was set level, and its surface was marked to show the Twenty-four Directions. The side of the upright meridian double ring was inscribed with the degrees of the celestial circumference. On the back of the fixed equator single ring were also inscribed degrees of celestial circumference. The north–south polar axis was set at an angle of 36^{d} to the horizon. These three components [horizon, meridian, and equator] were fixed and immovable. Their circumference was over 12 feet (*ch'ih* 尺).

This first array was called the Six Cardinal Points Component.

Threaded within this set of rings, and attached to the polar axle, were the (moveable) solstitial-colure double ring, the equator single ring, and the ecliptic ring, the latter intersecting the equator at an angle of 24^{d}. The side of the ecliptic single ring was inscribed with the twelve Jupiter Stations and the twenty-four Fortnightly Periods. By means of (a mechanism of) iron and silk thread, a sun-carriage travelled around the perimeter of the ecliptic ring, to distinguish between the times of day and night. These three rings were fixed together and rotated from east to west.

Within the ecliptic ring was attached the lunar single ring; it was set at an angle of slightly less than 6^{d} to the north and south of the ecliptic ring. Innermost was a single ring attached to the polar axle and intersecting the lunar ring at right angles. It engaged and propelled the moon-carriage [cf. Fig. 4.19], which moved to the east and duplicated the phases of the moon.

This second array was called the Three Arrangers of Time Component.

From the south-polar pivot there projected an iron rod shaped like a fork or a claw which held a land map [perhaps a globe?]. Below the horizon ring at the Four Corners [i.e. NE, SE, SW, NW], wooden dragon-shaped pillars held up the instrument.

[45] For the term 'Seven Regulators', i.e. the sun, moon, and five visible planets, see Ch. 2, n. 11 above.

[46] CMP, 3: 3b, 6a. An Chungt'ae travelled to China in 1733 and 1735 to investigate horological matters (Dr Gari Ledyard, private comm.).

Except for its larger size, the armillary sphere described in King Yŏngjo's essay is virtually identical to that on the surviving clock of Song Iyŏng (see Figs. 4.11–19 below). Because the armillary sphere described in this essay is clearly attributable to Yi Minch'ŏl (either by his own hand, or as an exact copy made by An Chungt'ae), we may safely conclude that the one on the surviving Song Iyŏng clock is also the authentic work of Yi Minch'ŏl.

The specified polar altitude of 36^d is of interest in the light of the charge given to Yi to model his sphere on 'Mr Ts'ai's commentary on the "Canon of Shun"'. The 'Mr Ts'ai' in question is most probably Ts'ai Shen 蔡沈; he was a pupil of Chu Hsi 朱熹, and his views would have carried great weight in Chosŏn, committed as that kingdom was to Neo-Confucian orthodoxy. Ts'ai Shen's commentary on the Book of Documents, the *Shu chi-chuan* 書集傳, was collected in the *Shu-chuan ta-ch'üan* 書傳大全 (Complete Commentaries on the Book of Documents; reprinted in Korea *c.* 1620) along with a diagram of an armillary sphere (see Fig. 4.10 below). We believe that Yi Minch'ŏl used that diagram as the basis for his own armillary sphere. The diagram specifies a polar elevation of 36^d, which is correct for the Northern Sung capital at Kaifeng. The surviving armillary sphere that we attribute to Yi Minch'ŏl, however, has a polar elevation which is correct for Seoul ($37° 41'$, or about 37.4^d). The latitude of Korea was known with fair accuracy at the time when King Yŏngjo wrote his essay; inscribed sundials of that period variously specify latitudes for Seoul of 37.2^d and 37.4^d. Evidently King Yŏngjo took his figure from the book, rather than from the instrument.

King Yŏngjo's essay goes on to describe the water-operated mechanism of the instrument in the Kyujŏng Pavilion as follows,

Chŭngbo munhŏn pigo, 3: 7a–b (1732):

The water-operation method (was as follows):

Alongside (the armillary sphere) the (jack-)mechanism wheels were held in a wooden framework 9 feet high and 5 feet broad, (that is,) about twice as high as it was broad. The armillary sphere was set up above a space south of the framework, its central axis going through the north pole into the framework, with a gear-wheel affixed. In the south-east corner was set a bronze water-vessel, with a clepsydra vessel as tall as the framework.

The essay goes on to describe a set of jackwork figures, all of the usual sort, set up in the south part of the framework to ring bells and hold tablets with the names of the double-hours. It then continues,

There were twenty-four metal balls, each about the size of a pigeon's egg. On the eastern side of the frame was an aperture leading to a passageway down which the balls ran at an angle.

Bronze was cast to make a float which was placed within the bronze inflow-vessel of the clepsydra. As water from the tank filled the vessel the float rose, and the wheels of the mechanism turned. On top of the whole apparatus the (Component of the) Three Arrangers of Time followed the motion of the heavens as if by itself. The metal balls fell one after another through their channels, propelling the jack-wheels. Every hour the bellman struck the bell and the time-annunciation officials popped up.

The drive-mechanism for Yi Minch'ŏl's instrument, as copied by An Chungt'ae, was thus presumably a clepsydra float with a ball-release mechanism similar to that of the old Striking Clepsydra (see Chapter 2, pp. 23–44). Except for a gear-wheel to drive the armillary sphere, the wheels referred to in King Yŏngjo's essay were simply carousels holding or operating the time-annunciating jacks. It turns out, then, that Song Chun'gil's 1669 petition requesting that the Jade Clepsydra be rebuilt did in fact result in the creation of new instruments very much in the tradition of the Jade Clepsydra and its companion, the Striking Clepsydra.

Yi Minch'ŏl's water-operated armillary clock (apparently his original instrument from the time of King Hyŏnjong) was repaired again in 1777,[47] and presumably both it and An Chungt'ae's copy of it were both still in operation for some years thereafter. These instruments were among the last survivors of the ancient Sino-Korean tradition of water-powered clocks. Their manner of representing the universe, and their modes of operation, were wholly traditional, but from the eighteenth century onwards clocks made under royal sponsorship in Korea had weight-drives of the sort pioneered by Song Iyŏng.

SONG IYŎNG'S WEIGHT-DRIVEN ARMILLARY CLOCK

Kim Sŏkchu's 1669 memoir, which gave us a good description of Yi Minch'ŏl's water-operated armillary clock, is much less informative about Song Iyŏng's instrument. It says only this,

Chŭngbo munhŏn pigo, 3: 3a (1669):

The armillary sphere set up by Song Iyŏng was in general design much the same as that of Yi Minch'ŏl, but instead of using water-tanks he used the gear-wheels of Western

[47] Jeon, STK, pp. 71–2.

clockwork. These mutually engaged, being made to appropriate sizes. The motions of the sun and moon, and the coming of dawn and dusk, and the announcing of the double-hours and intervals (*k'o*) were all without any error.

Kim Sŏkchu's description of two armillary clocks of generally similar design, one operated by a traditional clepsydral mechanism and the other by a Western weight-drive but both employing Sino-Arabic jackwork carousel-wheels, depicts the Yi court at the very moment of incorporating Western technology into its traditional astronomical instrumentation.

Western-style clocks were not entirely unknown in Korea before the time of Song Iyŏng. A striking clock of some kind was sent from Korea to Japan in 1607, when the Yi court was beginning to resume relations with Hideyoshi's successors, the Tokugawa Shogunate. Jeon believes that this instrument was a Western-style clock.[48] Chong Tuwŏn 鄭斗源, who introduced Korea's first telescope from Ming China in 1631, brought a Western clock back with him at the same time. That instrument was mentioned by Kim Yuk 金堉, along with another that Kim had seen in Peking in 1636; in his admiring comments, he acknowledged that he did not have any idea how the mechanisms worked.[49]

Knowledge of Western-style clockwork also reached Korea via Japan. Western weight-driven clocks were introduced into Japan as early as 1551[50] and began to be copied there by the early seventeenth century.[51] Presumably the earliest Japanese weight-driven clock mechanisms were direct copies of European ones, but during the seventeenth century modifications were introduced to suit the seasonally variable Japanese timekeeping system of six equal day double-hours and six equal night double-hours, somewhat analogous to the ancient East Asian system of five equal but seasonally variable night-watches running from 'dusk' to 'dawn'. (A similar system of variable hours was in general use in the West, except for astronomical purposes, until the advent of mechanical clockwork.) A modification which did not involve interference with the going rate of the clock

[48] Jeon, STK, p. 162; Yamaguchi Ryūji 山口隆二, *Nihon no tokei* 日本の時計 (The clocks of Japan) (Tokyo, 1942), pp. 17–18.

[49] Kim Yuk, *Chamgok p'iltam* 潛谷筆談 (Miscellaneous writings of Chamgok [= Kim Yuk]), quoted in Hong Isŏp 洪以燮, *Chosŏn kwahaksa* 朝鮮科學史 (History of Korean science) (Seoul, 1946), p. 160; see also Jeon, STK, p. 163.

[50] Nakayama, *History of Japanese Astronomy*, p. 123; Yamaguchi Ryūji, *Nihon no tokei*, p. 11ff; J. Drummond Robertson, 'The Clocks of Japan', in his *The Evolution of Clockwork* (London, 1931; repr. Wakefield: S. R. Publishers, 1972), pp. 189–287; see esp. pp. 195–7.

[51] Robertson, pp. 217–19; also T. O. R(obinson), 'A Transitional Japanese Clock', *Antiquarian Horology*, 1966, 5: 97.

mechanism was the introduction of a twenty-four-hour revolving dial carrying circumferentially adjustable plates to designate the double-hours and (in conjunction with supplementary plates at the half-double-hours where necessary) to trigger the striking mechanism if provided. Other methods depended upon retention of verge-and-foliot escapements, which had by this time largely been superseded in Europe by verge-and-balance-wheel escapements, but which in Japan facilitated twice-daily manual, or later automatic, adjustments of rate by amounts which were changed fortnightly according to the seasons,[52] just as the ball-racks in King Sejong's Striking Clepsydra had been changed every fifteen days to follow the variation in the length of the night-watches (see Chapter 2, pp. 27, 37). Another distinctive feature of Japanese clocks that is of interest to us here is their striking sequence. In early examples, the striking train ran through two sequences of six double-hours in each day-and-night period; these were indicated by a descending series of nine, eight, seven, six, five, and four bell-strokes for the double-hours, and a single stroke at each half-double-hour. In later examples, half-double-hours which followed double-hours indicated by an even number of bell-strokes were indicated by two strokes, and the others by a single stroke.[53] The striking train was let off by means of projections from the backs of circumferentially adjustable double-hour and half-double-hour plates on a revolving twenty-four-hour dial. In larger clocks this may well have presented frictional problems and thus led to the use of metal balls as power-relay devices (see Chapter 4, pp. 125–6 below).

During the reign of King Hyojong, that is, some time in the 1650s, a Japanese striking clock was brought to Korea where it was studied in detail by Yu Hŭngbal 劉興發. We learn of this only at second hand through the writings of Kim Yuk;[54] Kim says that Yu Hŭngbal understood the mechanism thoroughly, but does not himself go into much detail about it. He does, however, specify that the clock had a striking sequence of nine, eight, seven, six, five, and four strokes for each series of six double-hours, with a single stroke for the half-double-hours. This is clear evidence that it was of Japanese, and not European, origin.

This brings us again to the time of Song Iyŏng, who in 1669 may have been the

[52] Robertson, pp. 239–46.

[53] *Ibid.* p. 244. In each sequence of six double-hours, the signal of nine bell-strokes indicated the double-hour that included the moment (at the midpoint of the double-hour) of midnight or noon, i.e. the hours denoted *tzu* 子 and *wu* 午 in the duodenary series of Earthly Branches.

[54] See n. 49 above.

first Korean to construct a clock incorporating a Western-style mechanism. Nevertheless, the exact nature of what he did remains uncertain. The historical records do not state unambiguously that Song himself manufactured a completely new weight-driven mechanism to power the armillary instrument that he made at this date.[55] Because the clock has exactly the striking sequence found in early Japanese clocks, and has, moreover, a bell and driving weights of typically Japanese design (see Chapter 4, pp. 120, 128, and Fig. 4.2 below), it seems more likely that Song acquired an existing weight-driven movement of Japanese manufacture and made the adaptations necessary to cause it to rotate Yi Minch'ŏl's armillary sphere and to operate a traditional Sino-Korean jackwork-derived visible time-annunciating system. Even this, of course, would have been a work of considerable technical ingenuity; to say that Song Iyŏng may have acquired, rather than made with his own hands, the weight-driven mechanism of his clock is not to challenge his status as Korea's first recorded modern clockmaker.

A further reason for thinking that the existing movement may have originated in Japan is that the going train itself has no visible means of letting off the striking train. This suggests that the movement may have originated as one in which the striking train was intended to be let off from circumferentially adjustable double-hour and half-double-hour plates on a twenty-four-hour revolving dial showing Japanese variable hours (see p. 112; also Chapter 4, p. 120 below).[56]

Records from the reign of King Sukchong show that in 1687 Song's clock was repaired by Yi Chinjŏng 李鎭精;[57] a conversion, or attempted conversion, of its armillary sphere to represent a rotating-earth system may have taken place at that time, or later.

A reproduction of a Chinese striking clock was made for the Korean Bureau of Astronomy in 1715, and another Western-style clock was brought to the Yi court by a Chinese embassy in 1723.[58] The noted mathematician and astronomer Hong Taeyong 洪大容 (1731–1783) built a private observatory in the early 1760s that included a weight-driven armillary sphere and a weight-driven striking clock, as well as a water-operated celestial globe; on the basis of fragmentary evidence, it

[55] Jeon, STK, p. 166, implies that Song did so, but we find this improbable.
[56] Compare Robertson, 'Clocks of Japan', Fig. 9.
[57] CMP, 3: 3a; *Sukchong sillok*, 19: 13a; Jeon, STK, p. 71.
[58] Jeon, STK, pp. 163–4.

appears that Hong's weight-driven instruments were European, rather than Japanese, in conception.[59]

Much work remains to be done on the later history of clocks and clockmaking in Korea, but that would take us beyond the historical period studied in this book. Meanwhile, one may consult Jeon Sang-woon's surveys of these later events.[60] At the same time, one must remember that clockmaking of any kind remained somewhat exceptional in Korea even towards the end of the Yi period;[61] the average Korean was less likely to tell the time by consulting a clock than by looking at that most typical of Korean timekeeping instruments, a sundial – perhaps a 'pocket' scaphe or string-gnomon model.[62]

Kim Sŏkchu's 1669 description of Song Iyŏng's armillary clock would have remained merely a tantalising glimpse at the process of the Westernisation of Korean timekeeping instruments were it not for the survival, against all odds, of that clock down to the present day. The continued existence of Song's clock, still nearly intact, has allowed us to undertake the detailed analysis that we present in the following chapter.

[59] *Ibid.* pp. 41–2. Jeon refers to *Tamhŏnso* 湛軒書 (Works of Tamhŏn [= Hong Taeyong]), *oejip* 外集, *kwon* 卷 3, 8; we have, however, been unable to locate the specific passage there to which he refers. Although we feel, for the reasons given above, that the clock now in the Koryŏ University Museum is clearly attributable to Song Iyŏng and Yi Minch'ŏl, it is nevertheless necessary to raise and comment upon the hypothetical possibility that it is instead the clock built by Hong Taeyong *c.* 1760 for his private observatory. Hong, however, had visited Peking even before building his observatory, and so his armillary clock is likely to have been derived from the later European-style weight-driven clocks that were widely known in eighteenth-century China, rather than using an earlier Japanese-style mechanism like that of the Koryŏ University Museum instrument. Hong was very much interested in European astronomy and astronomical instruments; he was back in Peking in 1766, when he held a series of discussions with the Jesuit astronomers von Hallerstein and Gogeisl; see below, Ch. 5, n. 51.

[60] Jeon, STK, p. 164; 'Yissi Chosŏn ŭi sigye chejak sogo', pp. 107–8.

[61] Jeon, STK, p. 164.

[62] *Ibid.* Fig. 1.12.

4

THE ARMILLARY CLOCK OF SONG IYŎNG AND YI MINCH'ŎL (1669)

As we have seen, an investigation of the Korean historical records contained in the *Chŭngbo munhŏn pigo*[1] leaves little doubt that the armillary clock preserved in the Koryŏ University Museum is that constructed in 1669 by Song Iyŏng, incorporating an armillary sphere of Yi Minch'ŏl. The fortunate survival of this instrument from the reign of King Hyŏnjong allows us to verify and supplement the historical account presented in the previous chapter.

A study carried out by John H. Combridge on the basis of detailed photographs of the instrument has revealed a number of previously unknown, and possibly unique, mechanical and astronomical features of the clock.[2] The results of that study are presented here, together with references to the traditional Chinese instruments that are among the ancestors of Yi Minch'ŏl's armillary sphere.

GENERAL DESCRIPTION (FIGS. 4.1–2)

For convenience, this description assumes that the clock is orientated with the polar axis of the armillary sphere aligned on the celestial poles. The principal dimensions of the clock case are as follows:

Length, N to S	3 ft 11 in (120 cm)
Breadth, E to W	1 ft 9 in (52 cm)
Height, excluding plinth	3 ft 2 in (98 cm)

Items added to the original case by the Koryŏ University Museum authorities are:

(i) A wooden plinth.
(ii) A glass enclosure for the armillary sphere. Three lower and two side wooden

[1] CMP, 3: 2a.
[2] This study was first reported in (T. O. Robinson), 'A Korean 17th Century Armillary Clock'.

frame members of this enclosure are visible in the photographs; these frame members, as well as the plinth, are distinguishable in the photographs from the original cabinet-work by their lighter colour.

(iii) A glass top, fitting of which has occasioned removal of the tenons, visible in Rufus's photograph,[3] for a missing wooden top.

Missing side panels have also been renewed, but these, along with the glass case, were removed during photography.

The clock case is of Korean cabinet-work construction, with metal reinforcements, some or all of which may be later additions. It may conveniently be considered as comprising three sections, as follows:

(i) A low square pedestal at the S end, carrying the armillary sphere.

(ii) A narrow central section, defined by upright panels in the E and W sides of the case, containing the weights and striking train.

(iii) The remaining northern portion of the case, containing in its upper half the going train, time-annunciator and strike-release mechanisms, and the bell. A rectangular recess in the centre of the E side has a curved slot in its floor through which a medallion bearing the Chinese character for each double-hour appeared at the moment when the striking train was released at the beginning of that double-hour.

GOING TRAIN (FIGS. 4.3–4)

The verge-and-bob-pendulum going train, Fig. 4.3, is about 10 in high. It has four horizontal arbors and one vertical arbor, in a vertical frame orientated N–S on a rectangular base frame:

(i) The main arbor A1 (Fig. 4.4) revolved three times a day. It extends outside the frame and carries at its S end a 12-leaved bevelled lantern-pinion A2 driving the armillary sphere. At its centre, also outside the frame, it carries a 3-armed plain wheel A3 to which the driving chain-wheel A4 is coupled by ratchet-wheel A5 and pawls A6. At its N end it carries the 64-toothed main wheel A7 driving the second arbor and having attached to it, by four riveted

[3] 'Astronomy in Korea', Fig. 26; Rufus and Lee, 'Marking Time in Korea', Fig. 2; Needham, Wang, and Price, HC, Fig. 59; Needham, SCC III, Fig. 179.

Fig. 4.1. Outside view of Song Iyŏng's armillary clock, from the east.
This view shows the clock as it was when presented to the Koryŏ University Museum by Mr Kim Sŏngsu, but after a plinth had been added by the museum authorities and preparations had been made for fixing a glass top and a glass enclosure for the armillary sphere. The time-annunciator recess can be seen on the right, with the medallion bearing the Chinese character for the double-hour *yin*, 3 to 5 a.m., visible.

studs, a 120-toothed ring A8 driving the time-annunciator wheel by means of a 360-toothed gear-ring B2. The main wheel has three parallel-sided arms.

(ii) The second arbor C1, rotated 24 times a day by an 8-leaved pinion C2, carries a 48-toothed second wheel C3 driving the third arbor. Its N end is pivoted in an inverted U-shaped bracket C4, of Japanese type, riveted inside the frame to bridge over the 120-toothed ring A8.

(iii) The third arbor D1, rotated 8 times an hour by a 6-leaved pinion D2, carries a 42-toothed third wheel D3 driving the fourth arbor.

Fig. 4.2. Outside view of Song Iyŏng's armillary clock from the west.
In this view all of the removable panels have been removed from the case to reveal the time-annunciator wheel, strike-release mechanism, striking train, bell, and weights. The time-annunciator medallions bearing the Chinese characters for the double-hours *yu*, 5 to 7 p.m., *hsü*, 7 to 9 p.m., and *hai*, 9 to 11 p.m., are visible.

(iv) The fourth arbor E1, rotated 56 times an hour by a 6-leaved pinion E2, carries a 36-toothed, 4-armed contrate-wheel E3 driving the crown-wheel arbor.

(v) A short vertical arbor F1, rotated 336 times an hour by a 6-leaved pinion F2, carries the 15-toothed crown-wheel F3, requiring 168 pendulum beats to the minute. Its upper end is pivoted in a C-shaped bracket F4 allowing clearance for the verge G1, and its lower end is pivoted in an L-shaped bracket F5 riveted inside the frame. The lower end of the arbor is supported by an end-

Fig. 4.3. Going train, viewed from the south-east.

The escapement is seen at the top of the gear-train, and the main horizontal gear-ring of the timekeeping wheel at the bottom. The going train is located in the upper south-central portion of the main (northern) section of the cabinet.

spring F6 provided with a screw F7 for adjusting the depth of engagement of the pallets G6.

The horizontal verge G1 has a knife-edge G2 at its N end supported in a V-block G3 on a bracket G4 outside the frame, in which there is a clearance-hole G5 for the verge G1. This arrangement is illustrated by Symonds[4] as typical of English verge-and-bob-pendulum escapements. The pendulum-bob is missing from its rod G7. No evidence of conversion from verge-and-foliot or verge-and-balance-wheel is visible in the photographs.

The crown-wheel F3 is four-armed, and the other wheels of the going train three-armed. In contrast, the wheels on the polar axis of the armillary sphere are solid discs and, like the bevelled lantern-pinion A2 which drives the sphere, they have a distinctly more primitive appearance than the wheels of the going and striking trains. The lantern-pinion A2 shows some evidence of experimentation to find a satisfactory angle of bevel.

There is no photographic evidence of any means, within the going train, for letting off the striking train. We suspect that the movement originally belonged to a Japanese clock with a rotating twenty-four-hour dial carrying circumferentially adjustable double-hour and half-double-four plates for time indication and letting off the striking train; see pp. 125–6 and Chapter 3, pp. 112–13.

WEIGHTS (FIGS. 4.1–2)

Each of the driving weights for the going and striking trains comprises a heavy Japanese-style[5] near-cylindrical outer casting, with a number of auxiliary weight-adjustment discs held inside by a hexagonal nut on an axial rod. Provision for weight adjustment in a pendulum clock is, of course, something of an anachronism. It suggests that the maker may have been more familiar with verge-and-balance-wheel escapements, with which such provision would have facilitated adjustment of the going rate.[6] The hexagonal nut, if original, argues against European manufacture; it may be noted, however, that there are similar nuts on the metal tie-rods of the clock case, which may well be later additions. The weights are carried by chains having alternate long and short links.

[4] R(obert) W. Symonds, *A History of English Clocks* (Harmondsworth and New York: Penguin, 1947), p. 38.

[5] Robertson, *The Evolution of Clockwork*, part 2, 'The Clocks of Japan', Fig. 19.

[6] See also Ch. 3, pp. 111–12 above.

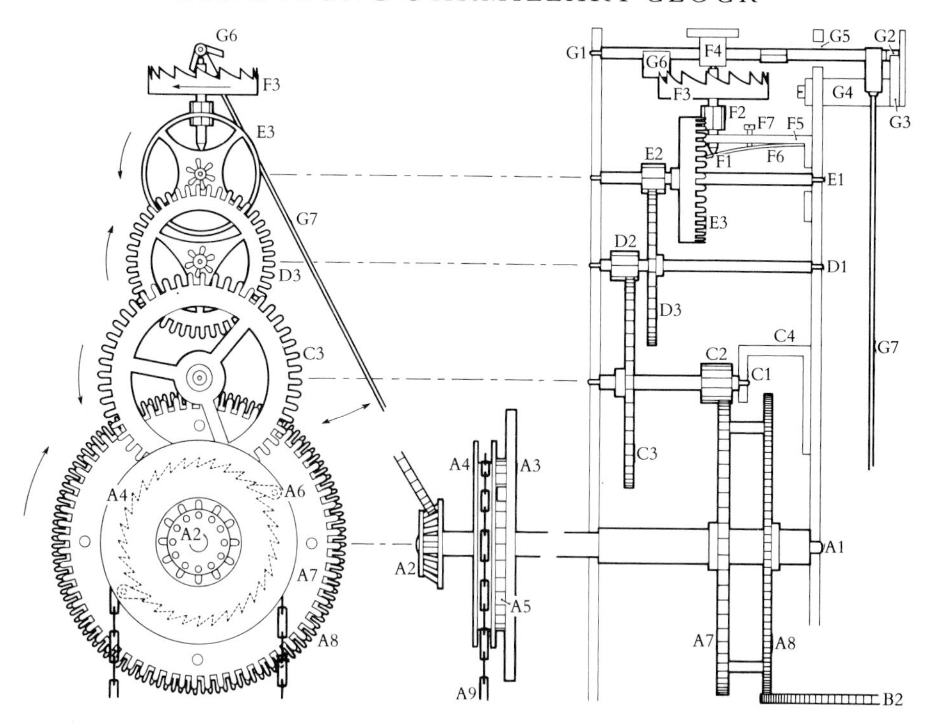

Fig. 4.4. Going train: explanatory drawing.

A1	Main arbor, rotating three times a day-and-night
A2	12-leaved bevelled lantern-pinion, driving 36-toothed wheel at north pole of armillary sphere
A3	3-armed plain wheel carrying ratchet-pawls A6
A4	Chain-wheel } coupled together, but free
A5	Ratchet-wheel } on arbor A1
A6	Ratchet-pawls, pivoted on A3
A7	64-toothed main wheel, driving C2
A8	120-toothed gear-ring, driving time-annunciator-wheel gear-ring B2
A9	Chain to weight
B2	360-toothed gear-ring of time-annunciator wheel
C1	Second arbor, rotating 24 times a day-and-night
C2	8-leaved pinion, driven by A7
C3	48-toothed second wheel, driving D2
C4	Bracket supporting second arbor
D1	Third arbor, rotating eight times an hour
D2	6-leaved pinion, driven by C3
D3	42-toothed third wheel, driving E2
E1	Fourth arbor, rotating 56 times an hour
E2	6-leaved pinion, driven by D3
E3	36-toothed contrate-wheel, driving F2
F1	Crown-wheel arbor, rotating 336 times an hour
F2	6-leaved pinion, driven by E3
F3	15-toothed crown-wheel
F4	Upper bearing bracket, with clearance for verge G1
F5	Lower bearing bracket, with clearance for arbor E1
F6	End-bearing spring for F1
F7	Screw for adjusting crown-wheel F3 to pallets G6
G1	Verge
G2	Knife-edge
G3	V-block bearing
G4	Bracket for G3
G5	Clearance-hole for verge G1
G6	Pallets
G7	Bob-pendulum rod, beating 168 to the minute

A long square weight H4 (Fig. 4.12), serving to tension the gut line H1, H2 for the annual-motion drive to the sidereal component of the armillary sphere, is concealed inside a narrow vertical chute H5 in the same part of the clock case as the driving weights.

TIME-ANNUNCIATOR WHEEL (FIGS. 4.5–6)

The time-annunciator wheel is constructed as a shallow cylindrical forged framework, rotated once a day on a square vertical arbor B1 (Fig. 4.6) in the centre of the clock case by a ring B2, of about 34 cm diameter, having 360 radial teeth which are appropriately formed on their upper edges for engagement in contrate-wheel fashion with the 120-toothed ring A8 (Fig. 4.4) on the main arbor of the going train. It carries twelve radial arms B3 (Fig. 4.6) pivoted at their inner ends, with freedom for their outer ends to rise and fall in slotted guides B4. Riveted nearly upright at the ends of the arms are medallions B5 bearing in relief the Chinese characters for the twelve double-hours, *tzu* 11 p.m. to 1 a.m., *ch'ou* 1 a.m. to 3 a.m., etc.[7] The medallion for the double-hour *shen*, 3 p.m. to 5 p.m., is now missing from its arm. As each arm was carried in turn from NNE towards ENE by the rotation of the time-annunciator wheel, its outer end slid up an inclined wire guide C towards the N end of the slot in the floor of the time-annunciator recess in the E side of the clock case (Fig. 4.1). At the instant of release of the striking train, at the beginning of a double-hour, the arm was lifted by the T-shaped top D1 of a vertical time-announcing rod D2, so that the medallion entered the slot from below and appeared in the annunciator recess (Fig. 4.1). The arm was retained in the lifted position by the horizontal upper portion of a metal guide-bar, the curved lower portions of which are visible in Figs. 4.1, 4.2, and 4.5, until the medallion approached the S end of the slot towards the end of the double-hour; it was then allowed to fall so that the medallion disappeared from view, perhaps synchronously with the appearance of the next medallion.

[7] Jeon (STK, pp. 87, 163) translates the (mostly untranslatable) Chinese 'Twelve Branches' names of the double-hours by the English names of the corresponding members of the Chinese 'animal cycle'. The duodenary Branch cycle and the associated Twelve Animals symbols can be used interchangeably to indicate the double-hours, as we saw in the case of King Sejong's scaphe sundials for public use (Ch. 2, no. 12, pp. 80–2 above), and also in that of the Jade Clepsydra. Sometimes the characters for the Branches are even pronounced as if they were the names of the corresponding animals; this usage is apparently common in Japan (see Robertson, *The Evolution of Clockwork*, pp. 198–208). But the use of animal symbols to designate the Twelve Double-hours has unfortunately sometimes been seen as lending support to the Western myth of a 'Chinese zodiac', on which see Combridge, 'Chinese Sexagenary Calendar-Cycles', p. 134.

Fig. 4.5. Time-annunciator wheel and strike-release mechanism, from the north-north-east.

The arm carrying the time-annunciator medallion bearing the Chinese character for the double-hour *ssu*, 9 to 11 a.m., is resting on the coat-hanger-shaped wire guide C (Fig. 4.6) leading upwards towards the time-annunciator slot; the medallions for *wu*, 11 a.m. to 1 p.m., and *wei*, 1 to 3 p.m., are also visible. A curved part of another guide, the horizontal top portion of which will hold each medallion in turn in the time-annunciator slot after the arm carrying it has been raised by the T-shaped top D1 of the vertical time-announcing rod D2, is visible in the gap in the cabinetwork near the left-hand end of the wire; another part of the same guide is visible in Figs. 4.1–2. In the foreground is a vase-shaped wooden hub H1 (Fig. 4.7) carrying at its far (i.e. south) end a small metal compartment-drum wheel H3 (more clearly visible in fig. 4.2) with nine shrouded vanes H4 for timing the strike-release balls, and at its near (i.e. north) end nine long radial arms H6 with bladed ends H7 for lifting the balls in turn after use.

STRIKE-RELEASE MECHANISM (FIGS. 4.1–2, 4.5, AND 4.7)

In the centre of the N end of the clock case a vase-shaped wooden hub H1 (Fig. 4.7) is mounted on a N–S horizontal metal arbor H2. The hub carries at its S end a small metal compartment-drum wheel H3 with nine shrouded vanes H4, and at its N end a metal disc H5 equipped with nine long radial arms H6 with square blades H7 riveted to their ends. The small compartment-drum wheel was rotated, anti-clockwise as seen from the N (Figs. 4.5 and 4.7), by the engagement of its vanes H4 with 24 pairs of downward-facing teeth J1 riveted to a ring J2 beneath the cylindrical frame carrying the time-announcing arms. Its function was to release at the beginning and middle of each double-hour one of a series of small metal balls, now lost, into a vertical chute K1 opening below the W side of a housing K2 surrounding the lower half of the compartment-drum wheel H3.

At the bottom of the vertical chute, the falling ball struck and depressed the N end of a strike-release treadle visible in Figs. 4.1 and 4.2, so causing the S end of this treadle to rise and release the striking train by means of a vertical strike-release rod visible in Figs. 4.2 and 4.8. At the same time, the N part of the strike-release treadle depressed the W end of the time-announcing treadle D3 (Fig. 4.6), so causing the E end D4 of this treadle to rise and at the beginning of a double-hour to lift one of the time-indicating medallions B5 into the time-annunciator recess by means of the T-shaped top D1 of the vertical time-announcing rod D2, as already explained.

After operating the strike-release treadle, the ball ran NW down an inclined runway M1 (Fig. 4.7) and entered the bottom of a near-vertical curved and slotted chute M2 in the NW corner of the clock case (Fig. 4.2). During the next double-hour, it was lifted up this chute by the blade H7 on one of the radical arms H6 at the N end of the wooden hub H1. At a point near the top of the chute M2 it entered a J-shaped inclined runway M3 along which it ran E and S to the E side of the housing K2 for the lower part of the small compartment-drum wheel, for reuse. In 1936 Rufus and Lee reported that the mechanism contained 'several iron balls'.[8]

The Chinese verge-and-foliot clock of Wang Cheng (1627) provides an apparent precedent for the use of metal balls in a weight-driven clock.[9] We now believe that the function of those balls was limited to sounding the drum and bell, either by direct percussion or by operating simple hammer-levers; and that the drawers near the base of the clock imply that they had to be returned to the top by hand for reuse, just as those in the Striking Clepsydra (Chapter 2, p. 33 above) had been.

[8] Rufus, 'Astronomy in Korea', p. 39; Rufus and Lee, 'Marking Time in Korea', p. 256.

[9] Needham, Wang, and Price, HC, pp. 146–7 and Fig. 53; Needham, SCC IV.2, pp. 513–15 and Fig. 669. The account of Wang Cheng's clock given in these works contains speculative interpretations which we no longer feel are justified. Wang Cheng's clock needs to be restudied in detail in the light of the new information that we have presented here.

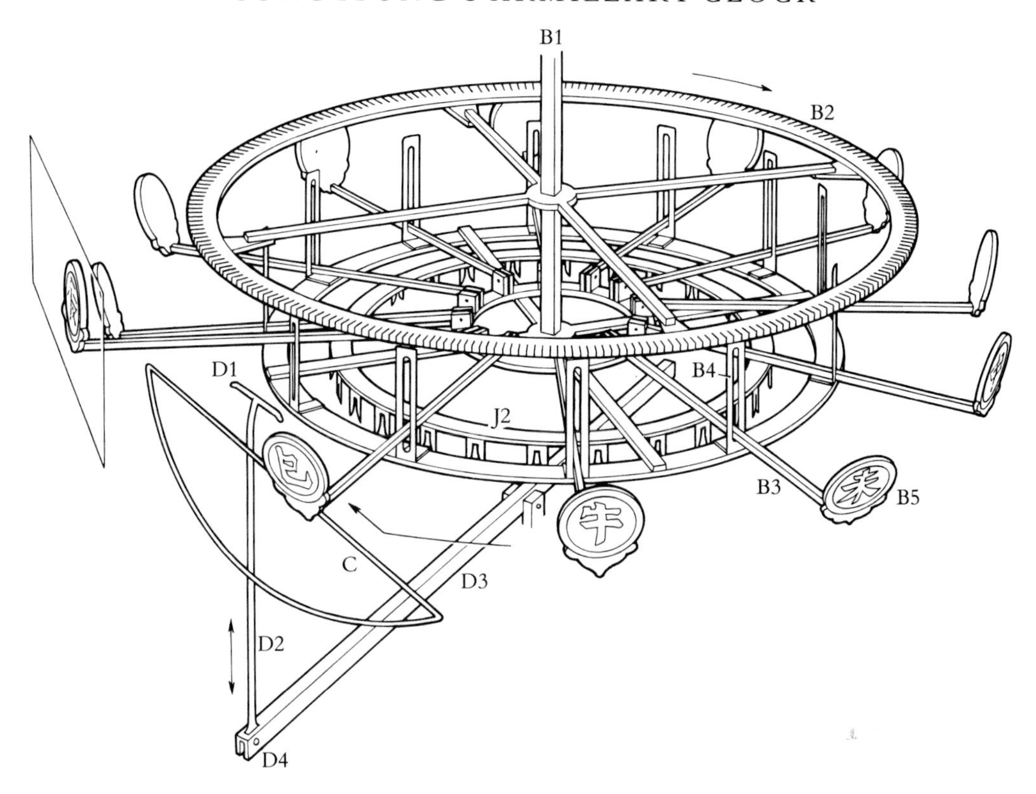

Fig. 4.6. Time-annunciator: explanatory drawing.

B1 Square vertical arbor, rotating once a solar day-and-night
B2 360-toothed gear-ring, driven by 120-toothed gear-ring A8 (Fig. 4.4)
B3 One of 12 radial time-announcing arms, pivoted at inner ends
B4 One of 12 vertical guides
B5 One of 12 time-announcing medallions

C Inclined wire guide, leading to the horizontal top of another guide not shown in this drawing but referred to in the caption to Fig. 4.5

D1 T-shaped top of D2
D2 Time-announcing rod
D3 Time-announcing treadle, operated by north end of strike-release treadle (visible in Figs. 4.1–2) every half-double-hour
D4 East end of D3

J2 Ring carrying 24 pairs of teeth J1 (Fig. 4.7)

The use of balls in the present clock for letting off the striking train can be viewed as an ingenious but elaborate means of avoiding the frictional problems liable to be encountered in direct letting off by the going train, especially in larger clocks. These problems would have been particularly troublesome in Japanese clocks of the type in which the striking train was let off by circumferentially adjustable double-hour (and when required, half-double-hour) plates on a rotating twenty-four-hour dial. If, as we suspect, the present movement was originally used in or intended for such a clock, this fact could well have led to the use of metal balls as power-relay devices, even before the addition of their other present function of operating a visible time-annunciator.

We wonder whether there were in East Asia – particularly in Japan – some weight-driven clocks of a hitherto unrecorded intermediate class (which may have included the present movement, in its original use or intended use) in which metal balls were used for letting off the striking train, but had to be raised by hand for reuse instead of being lifted automatically by the going train as they now are in the present clock. As we saw in Chapter 2 (pp. 41–4), the use of metal balls as a power-transmission device in clepsydras had a long history in East Asia. In the case of the present clock, one might regard the use of such balls as an instance of traditional East Asian technology being adapted to overcome a problem perceived in a newly imported European technology.

STRIKING TRAIN (FIGS. 4.8–9)

The striking train, Fig. 4.8, has five arbors pivoted E–W in a vertical frame which is similar to that of the going train, but has auxiliary plates riveted obliquely to it for the pivots of the two arbors F3, L7 (Fig. 4.9) carrying the levers F1, F2, L6 for locking and letting off the striking train.

(i) The main arbor A1 carries the 72-toothed main wheel A2, to which the chain-wheel A3 is coupled by ratchet-wheel A4 and pawls A5.

(ii) The second arbor B1 is driven from the main wheel by an 8-leaved pinion B2, and carries the 56-toothed pin-wheel B3, which has eight pins B4 riveted parallel to the arbor for operating the bell-hammer (visible in Fig. 4.2). Outside the W side of the frame, a count-pinion B5 with eight ratchet-shaped leaves for driving the count-wheel is pinned on a square B6 at the end of the arbor B1.

(iii) The third arbor C1 is driven from the 56-toothed pin-wheel B3 by means of a 7-leaved pinion C2 and carries the 54-toothed locking-hoop wheel C3. Outside the frame, the 45-ratchet-toothed Dutch-sixteenth-century-pattern count-wheel C5 is free to turn on the end of this arbor, where it is retained by a pin C6 and 3-legged spring C7.

(iv) The fourth arbor D1 is driven from the 54-toothed locking-hoop wheel C3

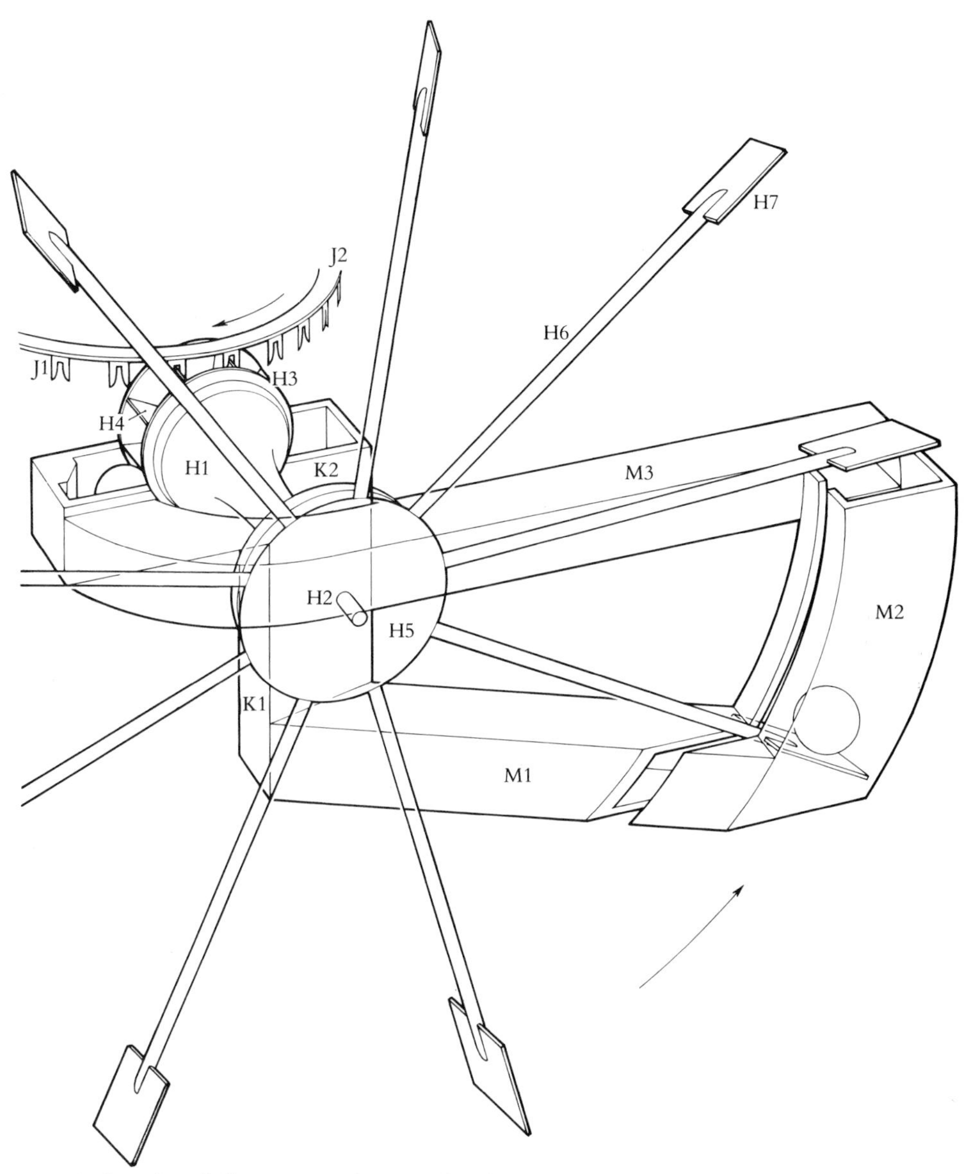

Fig. 4.7. Strike-release ball-system: explanatory drawing.

H1	Wooden hub	J1	One of 24 pairs of teeth for rotating H3
H2	Metal arbor, rotating $2\frac{2}{3}$ turns a day-and-night	J2	Ring carrying 24 pairs of teeth J1
H3	Nine-compartment ball-timing compartment-drum wheel	K1	Ball-drop chute
H4	One of nine vanes of H3	K2	Housing for ball-timing compartment-drum wheel H3
H5	Metal disc	M1	Ball-return runway
H6	One of nine ball-lift arms	M2	Ball-lift chute
H7	One of nine ball-lift blades	M3	Ball-supply runway

by a 6-leaved pinion D2, and carries a 42-toothed fourth wheel D3 driving the fifth arbor.

(v) The fifth arbor E1 is driven from the 42-toothed fourth wheel D3 by a 6-leaved pinion E2, and carries a double-vaned European-type fly E3 with a friction grip on the arbor.

The count-wheel C5 still bears the maker's setting-out lines. It is designed so that the hourly striking sequence during each twelve hours – six double-hours – was 9, 1, 8, 1, 7, 1, 6, 1, 5, 1, 4, 1. This is one variant of the Japanese striking sequence, in another, and later, variant of which the hours following the odd-numbered double-hours were indicated by single strokes, and those following the even-numbered double-hours by two strokes.[10] Its introduction into Korea during the reign of King Hyojong (1650–9) is referred to in an essay by Kim Yuk, who in 1645 was head of the Korean astronomical board.[11]

The bottom of the vertical strike-release rod may be seen in Fig. 4.2 and its top in Fig. 4.8. All the wheels of the striking train, except the count-wheel, are 3-armed. As in the going train, the lower wheels have straight arms and the upper wheels have arms of curved outline. The bell, visible in Fig. 4.2, is of a characteristically Japanese shape.[12]

THE ARMILLARY SPHERE (FIGS. 4.11–20)

The demonstrational armillary sphere embodies many features of the classical and medieval Chinese armillary spheres,[13] together with others showing the effect of European influence via China or Japan. Mechanically, it is a good deal more elaborate than the early-nineteenth-century Chinese spring-driven clockwork celestial globes now in the Guildhall Library Museum of the Worshipful Company of Clockmakers of the City of London and elsewhere.[14] It originally

[10] Robertson, *The Evolution of Clockwork*, pp. 197, 200, 244; Jeon, STK, p. 163; F. A. B. Ward, *Time Measurement, Part I* (1st edn, 1936), p. 43. The sequence in which each double-hour is followed by a single stroke at the half-double-hour is earlier than the variant employing alternate double strokes, which tends to confirm the early date of the mechanism of this clock.

[11] Rufus, 'Astronomy in Korea', p. 37.

[12] Robertson, *The Evolution of Clockwork*, part 2, p. 223 and Figs. 4, 7, 12, 14.

[13] Needham, SCC III: 339ff and Table 31.

[14] H. L. Nelthropp, *Catalogue of the Nelthropp Collection* (2nd edn, London, 1900), p. 10, Cat. no. 25; C. Clutton and G. Daniels, *Clocks and Watches in the Collection of the Worshipful Company of Clockmakers* (London, 1975), p. 96, Cat. no. 590 (illustrated).

The Clockmakers' Company's globe is inscribed with its maker's name Ch'i Mei-lu 齊梅麓 of Wuyuan

Fig. 4.8. Striking train, viewed from the south-south-east.
The upper part of the strike-release rod may be seen to the right; its lower part is visible in Fig. 4.2. The location of the striking train relative to the mechanism as a whole may be seen in Figs. 4.2 and 4.11.

comprised no fewer than five component stages or layers:

(i) A fixed outer terrestrial-coordinate component B (Figs. 4.14, 4.17).

(ii) A solar component C, rotated about a fixed polar axis once a day by the clock, to drive a sun-model C9 (Fig. 4.16) along a shallow spiral trace corresponding to the apparent annual path of the sun in the heavens. This component is now missing, but the carriage C7 and supporting peg C8 for the sun-model, and solar-time pipes and wheels at the polar pivots, remain.

(iii) A sidereal component D (Figs. 4.14, 4.17), with equator, ecliptic, and lunar path, coupled to the solar component by an annual-motion compensating drive so as to rotate about the fixed polar axis once more than the solar component in the course of a year, thus keeping sidereal time in correspondence with the apparent rotation of the celestial sphere around the earth.

(iv) A lunar component E, driven from the solar component by a lunar-motion train (Fig. 4.17) so as to rotate $28\frac{1}{2}$ times about the fixed polar axis for every $29\frac{1}{2}$ rotations of the solar component. This component drove a moon-model, now missing, around the lunar-path track of the sidereal component once each month, and in the annual course of this movement a phase-of-moon mechanism, now missing, was operated through approximately $12\frac{1}{2}$ lunar cycles each year.

(Kiangsi, formerly Anhui) and a date equivalent to June/July 1828. It has been identified by one of us (JHC, unpublished) as one of a series by this maker and others. The name Ch'i Mei-lu, which is the courtesy-name of the noted provincial official Ch'i Yen-huai 齊彥槐, also appears on the similar globe, dated April/May 1830, formerly in the Anhui Provincial Museum but now on loan to the National Historical Museum in Peking (SCC IV.2: 527–8 and Fig. 670). A third example bearing a variant form of the same courtesy-name, and dated March/April 1830, was sold in New York (Sotheby Parke Bernet, Inc., *Watches, Scientific Instruments* ..., June 14, 1928, Cat. no. 210, illustrated). A fourth example bearing two other names, and dated September 1830, was sold in Paris, with a modern replacement stand (H. Chayette, *Montres et pendules de collection*, Paris, 24 novembre 1980, Cat. no. 154, illustrated). A fifth, similar but not identical, and unattributed, example is in the antique instrument collection of the Adler Planetarium, Chicago. The dismounted shell of a sixth example, also unattributed, was acquired by the Royal Scottish Museum, Edinburgh, in 1984 (Inv. no. RSM TY 1984.102).

The globes are engraved with reversed star-maps, closely resembling those of twenty-four half-gores published by Hsü Ch'ao-chün 徐朝俊 in part 4.1, 'T'ien-ti t'u-i' 天地圖儀, of his *Kao-hou meng-ch'iu* 高厚蒙求 (SCC III: 456 and IV.2: 531, n. d). In the Chicago and Edinburgh examples the stars have individual numbers corresponding to those of Hsu's large double-planispheric ecliptic star-map *Huang-tao Chung-hsi ho-t'u* 黃道中西合圖, 1807 (see Alexander Wylie, *Notes on Chinese Literature* (2nd edn, Shanghai, 1902), p. 124, and compare the star-numbers in Wylie's 'List of Fixed Stars', c. 1850, reprinted in his *Chinese Researches*, part III, pp. 110ff), of which there is an example in the library of the School of Oriental and African Studies, University of London. The ecliptic circles of the globes are engraved with the sun's position at the 24 *ch'i*, subdivided to five-day groups of single days. The globes are made to rotate diurnally, by internal clock movements having spring drives, fusees with chains, and verge escapements.

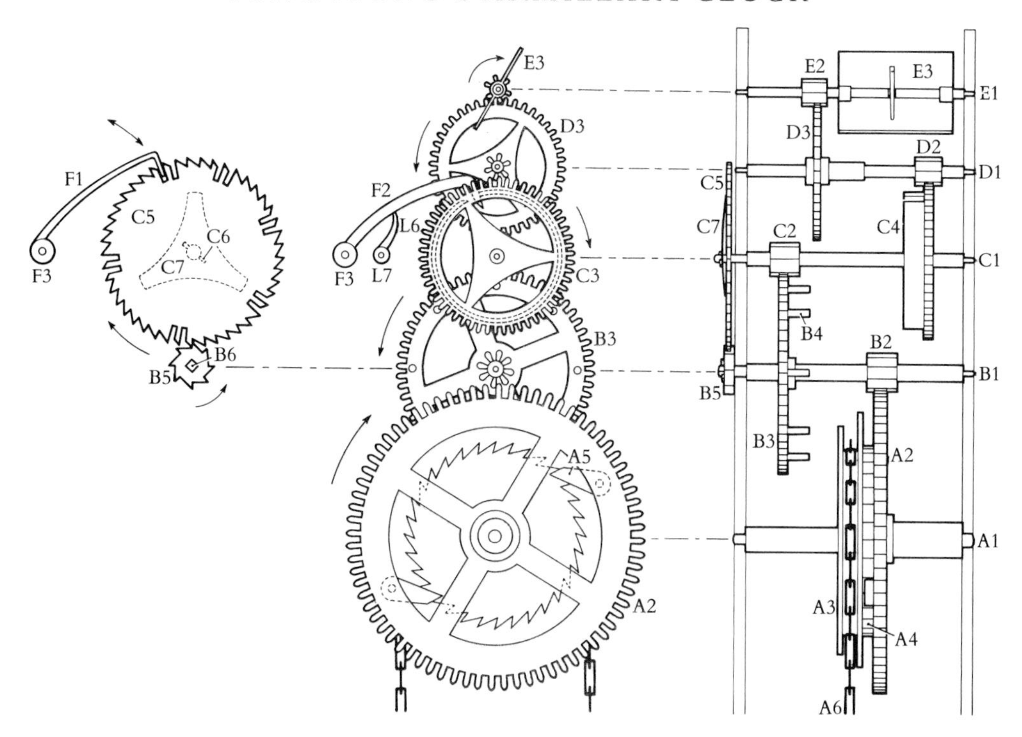

Fig. 4.9. Striking train: explanatory drawing.

A1 Main arbor
A2 72-toothed main wheel
A3 Chain-wheel } coupled together, but free
A4 Ratchet-wheel } on arbor A1
A5 Ratchet-pawls, pivoted on A2
A6 Chain to weight

B1 Second arbor
B2 8-leaved pinion, driven by A2
B3 56-toothed pin-wheel, driving C2
B4 One of eight hammer-lifting pins
B5 8-leaved pinion, driving count-wheel C5
B6 Square seating for pinion B5

C1 Third arbor
C2 7-leaved pinion, driven by B3
C3 54-toothed locking-hoop wheel, driving D2
C4 Locking hoop
C5 45-toothed count-wheel, free on arbor C1 to allow rotation by B5 and resetting when necessary
C6 Pin
C7 Three-legged retaining spring for count-wheel C5

D1 Fourth arbor
D2 6-leaved pinion, driven by C3
D3 42-toothed fourth wheel, driving E2

E1 Fifth arbor
E2 6-leaved pinion, driven by D3
E3 Fly, with friction grip on arbor E1

F1 Count-lever } on a common arbor F3; see
F2 Locking lever } Fig. 4.8
F3 Arbor carrying count-lever F1 and locking lever F2

L6 Letting-off lever
L7 Arbor carrying letting-off lever L6 and the strike-release lever visible in Fig. 4.8

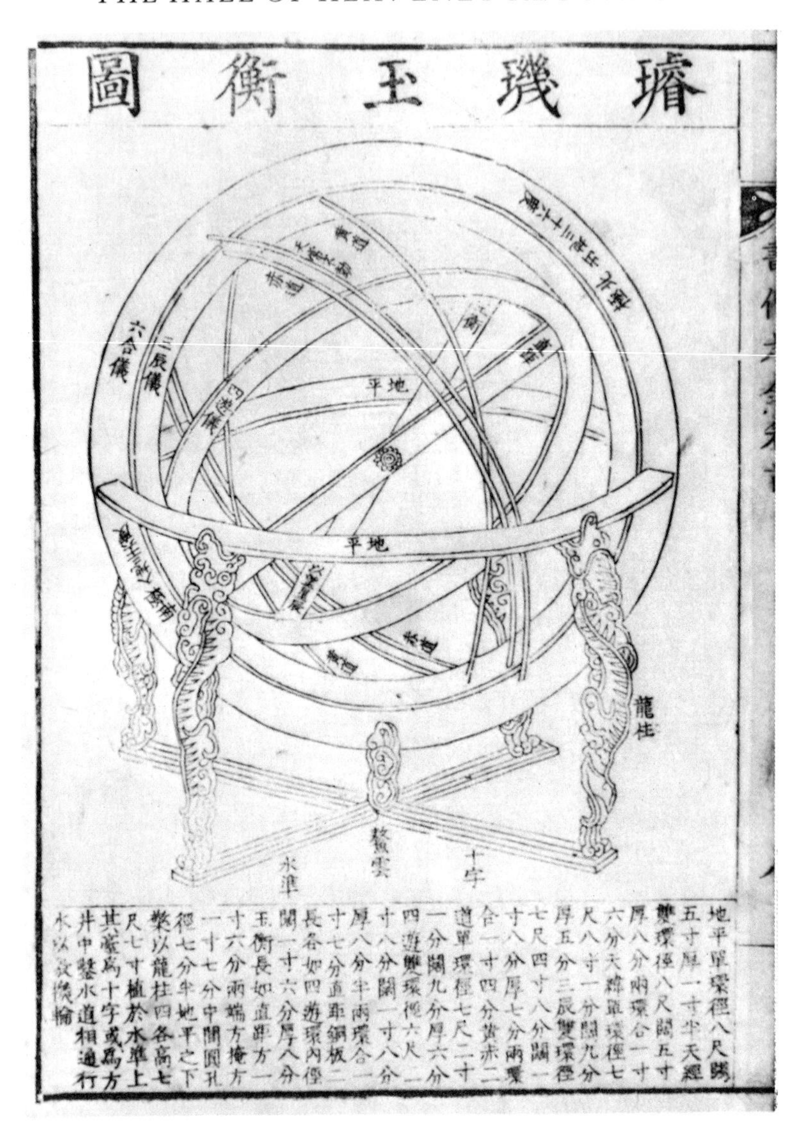

Fig. 4.10. A late representation of a medieval Chinese observational armillary sphere.

This drawing, which we reproduce from the *c.* 1620 Korean edition of *Shu-chuan ta-ch'üan* (Complete Commentaries on the Book of Documents), derives ultimately from an illustration in Yang Chia's *Liu-ching t'u* (Illustrations of the Six Classics) of about 1160 C.E. It was probably used by Yi Minch'ŏl when designing the armillary sphere for the present clock. The title of the drawing is 'Hsüan-chi yü-heng t'u' (Illustration of an armillary sphere), and the text gives the diameters of the outer rings as 8 feet. (See above, Chapter 2, n. 11, and Chapter 3, n. 34, on the terms *hsüan-chi* and *yü-heng* and their later extension to refer (as here) to armillary spheres.) The polar altitude stated on the meridian ring, 36 Chinese celestial degrees, is that of the Northern Sung capital at Kaifeng, from which all the astronomical instruments were carried away in 1126 C.E. by the Jurchen Tartars (Needham, Wang, and Price, HC, pp. 132, 134).

Fig. 4.11. Yi Minch'ŏl's demonstrational armillary sphere, viewed from above, with north at top of picture.
This view shows all the rings of the armillary sphere, prominent among which are the ecliptic ring D5 (Fig. 4.12) with its sun-path channel D6 (Fig. 4.16) and projecting square sun-carriage peg C8, and the lunar-path ring D7 (Fig. 4.19) with its 27 axially aligned phase-of-moon actuating pegs D8. At the north polar pivot may be seen the 36-toothed solar-time wheel C2 (Fig. 4.12) and 12-leaved pinion C4, and the 48-toothed annual-motion contrate-wheel G2. Part of the gut line H2 for the annual-motion drive may be seen emerging from the north polar pivot and wrapped round the slender central portion of the contrate-wheel arbor G1, and another part emerging from its conduit H3 above and to the right of the pivot, and entering the chute for the square tension-weight H4.

(v) A fixed inner component, consisting of a terrestrial globe (Fig. 4.20), with meridians and parallels together with geographical markings and names in Chinese characters, supported on a stationary polar axis F1 (Figs. 4.14, 4.17).[15]

Armillary fixed outer component

This component corresponds to the 'Six Directions Instrument' of the classical and medieval Chinese armillary spheres.[16] It comprises one double and two single bronze rings:

(i) The terrestrial horizon single ring B1 (Fig. 4.12), diameter 41.3 cm, is engraved on its upper surface with the Chinese characters for the Twenty-four Directions,[17] and is carried by a wooden supporting ring on four

Fig. 4.12. Yi Minch'ŏl's demonstrational armillary sphere: explanatory drawing.

B1 Fixed outer terrestrial-horizon single ring
B2 Meridian double ring
B3 Fixed equator single ring

C2 36-toothed wheel
C3 Outer, solar-time, pipe
C4 12-leaved pinion
C6 57-toothed solar-time gear-wheel
C7 Sun-carriage, sliding within sun-path channel D6 (Fig. 4.16) mounted on outer edge of ecliptic ring D5
C8 Square supporting peg for sun-image C9 (Fig. 4.16)

D1 Solstitial-colure double ring
D2 Hollow upper quadrant of one element of D1
D3 Inner, sidereal-time, pipe
D4 Revolving equator single ring
D5 Ecliptic single ring, with sun-path channel D6 (Fig. 4.16) mounted on its outer edge
D7 Moon-path single ring, carrying moon-carriage ring E4 (Fig. 4.19)

E1 Moon-transport single ring
E3 59-toothed lunar-motion gear-wheel

F Terrestrial globe

G1 Arbor of annual-motion-work contrate-wheel G2, with gut line H1, H2 encircling its reduced-diameter middle section
G2 48-toothed contrate-wheel, rotating once in four days-and-nights

H1 Gut line from sun-carriage C7 via sun-path channel D6 (Fig. 4.16), D2, and D3
H2 Gut line H1 continuing via H3 to tension-weight H4
H3 Metal tube, housing gut line, concealed in wooden frame-upright
H4 Gut-line tension-weight, in square-section wooden chute
H5 Square-section wooden chute for H4

[15] Jeon (STK, pp. 19, 21, 70) suggests that the polar axis was designed to rotate diurnally, but we find no evidence to support that view. The present study leaves no doubt that the armillary sphere originally had a stationary earth-model surrounded by rotating lunar, sidereal, and solar components in accordance with Chang Heng's *hun t'ien* system (STK, pp. 13–15). It may have been altered later; see nn. 31 and 38 below.

[16] Needham, SCC III: 339ff and Table 31; cf. H. Maspero, 'Les instruments astronomiques des chinois aux temps des Han', p. 309.

[17] Needham, SCC III: 576; IV.1: 297 and Table 51.

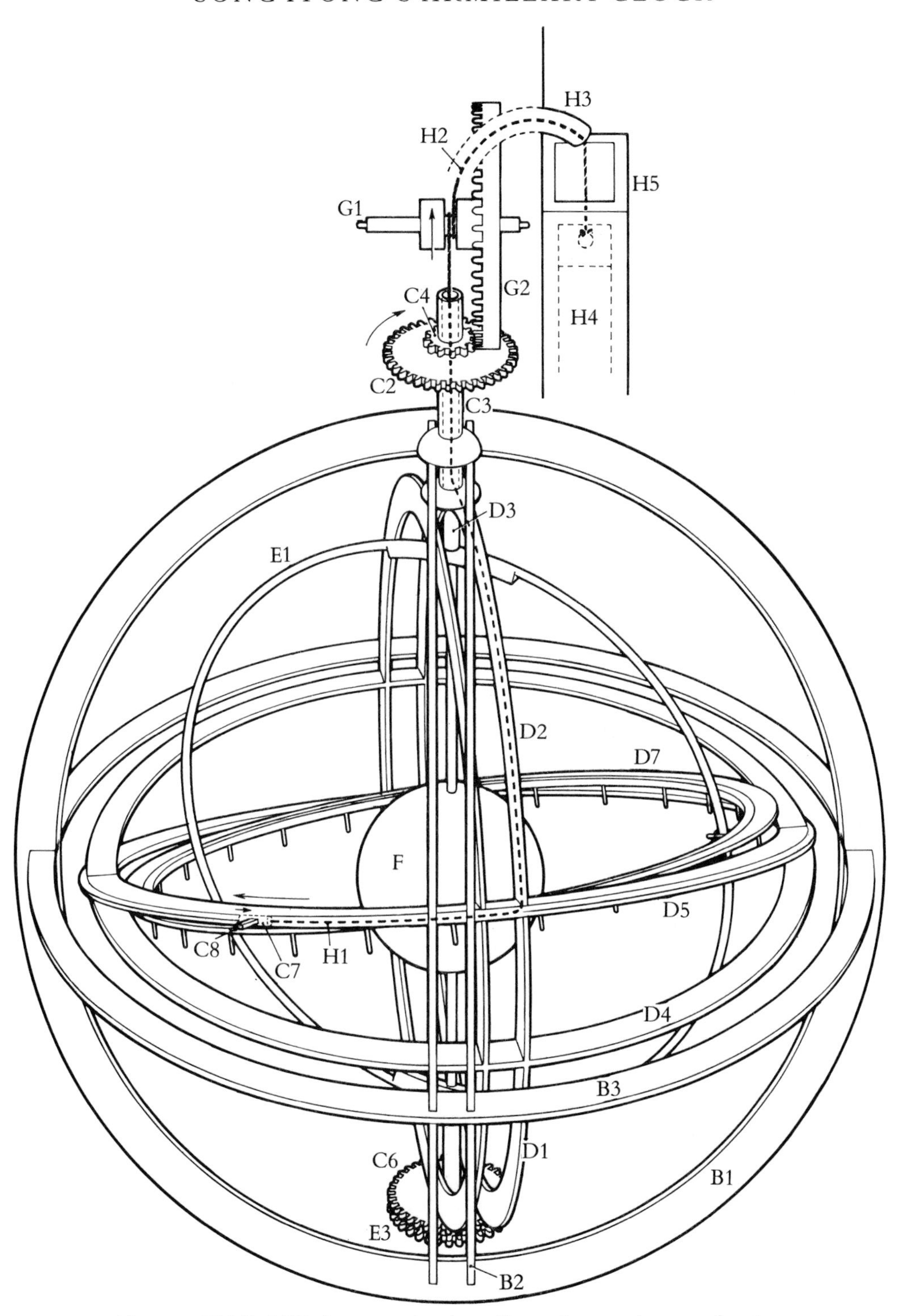

Fig. 4.12. Yi Minch'ŏl's demonstrational armillary sphere: explanatory drawing.

wooden conventionalised dragon-columns at the 'Four Corners of the Earth', NE, NW, SE, SW.

(ii) The meridian double ring B2, diameter about 39 cm, intersects the N and S horizon-points perpendicularly, and carries the N and S polar pivots for the revolving components at a polar altitude corresponding to the latitude of Seoul, 37° 41′. Its lowest point is supported centrally by a bronze conventionalised tortoise-and-cloud column.[18]

(iii) The fixed equator single ring B3, diameter about 40 cm, intersects the meridian double ring B2 at right angles, and the E and W horizon-points at an angle of about $52\frac{1}{2}°$. It is engraved near the inner edges of its N and S polar faces with 360 European degrees, grouped in 10s and 30s by longer graduations.

The wooden dragon and bronze tortoise-and-cloud column supports of the armillary sphere rest on an X-shaped wooden base, which is flush with but structurally distinct from the rest of the top surface of the clock-case pedestal. This base corresponds to the 'X-shaped water-level base'[19] used with the classical and medieval Chinese armillary spheres down to at least Sung times, but superseded by a square water-level base in Kuo Shou-ching's armillary sphere of about 1276 C.E.[20]

The four dragon-columns which support the horizon ring are of the much conventionalised form seen in late derivatives from Yang Chia's *Liu-ching t'u* (Illustrations of the Six Classics) of about 1160 C.E.,[21] from which they have often been reproduced in Western as well as Chinese publications.[22] The *Chŭngbo munhŏn pigo* records that in 1669 Yi Minch'ŏl was ordered to cast a bronze armillary sphere 'on the basis of Mr Ts'ai's commentary on the "Shun-tien" chapter of the *Shu ching*'.[23] The original edition of Ts'ai Shen's book was printed

[18] Maspero, 'Les instruments astronomiques', p. 320.

[19] *Hsin i-hsiang fa-yao*, 1: 20b; cf. Maspero, 'Les instruments astronomiques', p. 319.

[20] Needham, SCC III: 367ff and Figs. 156 and 163; Edward L. Stevenson, *Terrestrial and Celestial Globes: Their History and Construction*, 2 vols. (New Haven: Yale University Press, 1921), II: Fig. 117.

[21] E.g. *Liu-ching t'u ting pen*, 1740 edn, pp. 14b–15b. Cf. a reference by Wylie in Sir Henry Yule, *The Book of Ser Marco Polo*, 2 vols. (3rd edn, ed. H. Cordier, London, 1903), II: 450–1, to a *Luh-king-too-kaou*, 'Illustrations and Investigations of the Six Classics'.

[22] J. P. G. Pauthier, *Chine (ancienne)*, premier partie (Paris, 1839), Pl. IV; H. Medhurst, *Ancient China: The Shoo King* (Shanghai, 1846), pl. facing p. 16; Needham, Wang, and Price, HC, Fig. 30 (from a copy inserted in an eighteenth-century MS. of *Hsin i-hsiang fa-yao* in the Peking National Library; Cullen and Farrer, 'On the Term *Hsüan Chi* and the Flanged Trilobate Jade Disc', Pl. II.

[23] CMP, 3: 2a; see Ch. 3, pp. 105 and 109 above.

in 1209 and has no illustrations; but his *Shu ching* commentary, with those of other writers and a selection of illustrations taken from Yang Chia's *Liu-ching t'u*, was later included in the *Shu-chuan ta-ch'üan* (Complete Commentaries on the Book of Documents), an illustrated edition of which was printed in Korea *c.* 1620. Fig. 4.10 is a reproduction of the illustration showing the armillary sphere in that edition, which it seems probable was the one used by Yi Minch'ŏl for the design of the present instrument.

The drawing, if not distorted during copying, was probably intended as no more than a rough sketch, since there is good evidence that the dragon supports of Sung and Yüan armillary spheres were actually modelled in very 'lifelike' detail. The best contemporary drawings are those of the dragons entwining the upright columnar supports of the horizon ring in the non-mechanised observational armillary sphere installed on the top of Chang Ssu-hsün's mercury-driven clock-tower of 979 C.E., which were printed in Shih Yüan-chih's 1172 edition of *Hsin i-hsiang fa-yao*.[24] The text of this work explains that while straight dragon-column supports had been used in the ancient model, in the *yüan-feng* reign-period (1078–85 C.E.) curved dragon supports had been found more convenient for the observers, and that this improvement had also been incorporated in the shaft-driven observational armillary sphere of 1088.[25] The actual appearance of the dragons at this, perhaps the highest, point of their evolution is not shown in the book, but can be appreciated from those in the non-mechanised armillary sphere of Kuo Shou-ching (1276 C.E.)[26] which was modelled after that of 1088, and a fifteenth-century replica of which still exists.[27]

The tortoise-and-cloud column central support has an even longer recorded history.[28] It was first used in the iron armillary sphere cast during the *yung-hsing* reign-period (409–14 C.E.) in the Wei Dynasty. In the clock-driven observational armillary sphere of the *yüan-feng* reign-period (1078–85 C.E.) a split tortoise-and-cloud column was used, to allow a central space for the driving chain.[29] A hollow

[24] *Hsin i-hsiang fa-yao*, I: 6a, 19a. Cf. Maspero, 'Les instruments astronomiques', Fig. 18; Needham, Wang, and Price, HC, Fig. 6; Combridge, 'The Astronomical Clocktowers of Chang Ssu-hsun and his Successors, A.D. 976 to 1126', Figs. 3 and 4.

[25] *Hsin i-hsiang fa-yao*, I: 19b; Maspero, 'Les instruments astronomiques', p. 320.

[26] S. W. Bushell, *Chinese Art* (London, 1909), I: 91; cf. Needham, Wang, and Price, HC, p. 133, n. 2; Needham, SCC IV.2: 408.

[27] Needham, SCC III, Figs. 156 and 163; Stevenson, *Globes*, II: Fig. 117; Yule, *Marco Polo*, II: 451; Bushell, *Chinese Art*, I: Fig. 64; Combridge, 'Astronomical Clocktowers', Fig. 18.

[28] *Hsin i-hsiang fa-yao*, I: 20a; cf. Maspero, 'Les instruments astronomiques', p. 320.

[29] *Hsin i-hsiang fa-yao*, I: 19b; Combridge, 'Astronomical Clocktowers', Figs. 11, 13.

cylindrical tortoise-and-cloud column was introduced to conceal the improved vertical-shaft drive of the 1088 C.E. armillary. It was this last hollow cylindrical tortoise-and-cloud column which Kuo Shou-ching copied in realistic detail in his non-mechanised observational armillary of 1276 C.E.[30] The column more conventionally represented in Yang Chia's drawing, and thence in its Korean miniature realisation, is of the shorter proportions appropriate to a large waist-level demonstrational armillary, as distinct from the 8-foot-diameter above-eye-level observational armillary referred to in the accompanying text.

Armillary solar component (now missing)[31]

This component formerly occupied the space now vacant between the fixed outer terrestrial-coordinate component B (Figs. 4.14, 4.17) and the revolving sidereal component D. The positions where it was fixed to the solar-time pipes C3, C5 at the N and S polar pivots are now occupied by bun-shaped Korean bronze spacing pieces (Figs. 4.1–2, 4.11, 4.13).

The component was rotated once a solar day by the engagement of the 12-leaved bevelled lantern-pinion A2 (Fig. 4.14), on the S end of the main arbor A1 of the going train, with a 36-toothed wheel C2 on the solar-time pipe C3 at the N polar pivot. Its function was to propel a sun-transport carriage C7 (Fig. 4.16), which still exists, along a sun-path channel D6 mounted on the outer edge of the ecliptic ring D5 of the sidereal component.

Its form was probably that of a polar-pivoted single ring C1 (Fig. 4.16), similar to the existing moon-transport ring E1 (Fig. 4.17) but divided on one side as shown at C1 in Fig. 4.16 to embrace the square peg C8 projecting from the sun-carriage C7 and to allow for its changes in declination during the year, in a way similar to that of the sun-transport rings on the European clock-driven celestial globes of Georg Roll, Isaac Habrecht III,[32] and others.[33]

The solar-time pipe C3 (Figs. 4.12, 4.14) at the N polar pivot carries above its 36-toothed driving wheel C2 a 12-leaved pinion C4 for the annual-motion drive.

[30] Needham, SCC III, Figs. 156 and 163; Stevenson, *Globes*, II: Fig. 117; Yule, *Marco Polo*, II: 451; Bushell, *Chinese Art*, I: Fig. 64; cf. Combridge, 'Astronomical Clocktowers', Fig. 18.

[31] The absence of the armillary solar component could be due to a conversion (or attempted conversion) of the instrument to a rotating-earth system; see n. 15 above and n. 38 below.

[32] Both represented by examples in the National Maritime Museum, Greenwich, England.

[33] Cf. Hans von Bertele, *Globes and Spheres* (Lausanne: Scriptar S.A., 1961), *passim*; Henry C. King, *Geared to the Stars* (Toronto, 1978), pp. 83–6, Figs. 5.22–4; Stevenson, *Globes*, I: Fig. 74.

Fig. 4.13. Detail of armillary-sphere north-polar clock-drives, viewed from the south-west.

The main arbor A1 of the going train (Fig. 4.4) carries at the left of this view the driving-chain wheel A4, and in the centre the 12-leaved bevelled lantern-pinion A2 for driving the 36-toothed solid gear-wheel C2 (Fig. 4.12) on the solar-time pipe C3 at the north-polar pivot of the armillary sphere. Above the 36-toothed gear-wheel C2 is the 12-leaved solid pinion C4 for driving the 48-toothed four-armed annual-motion contrate-wheel G2. The horizontal arbor G1 of the contrate-wheel G2 revolved once in four days-and-nights, and the diameter of its slender central part is such that the gut line H2 wrapped round it was pulled out from the north-polar pipe D3 through a length equal each day-and-night to that of one Chinese celestial degree on the ecliptic ring. The top of the vertical chute for the gut-line tension-weight H4 is behind the contrate-wheel G2, and the filling of the curved channel in which the metal tube H3 for the gut line is concealed can be seen in the front surface of the central timber upright. The bun-shaped spacer between the outer fixed meridian double ring B2 and the intermediate solstitial-colure double ring D1 occupies the position of the missing sun-transport single ring C1 (Fig. 4.16). The inner moon-transport single ring E1 (Fig. 4.12) is free to turn around the protruding lower end of the inner sidereal-time pipe D3 (Fig. 4.14), which also provides a support for the upper end of the terrestrial-globe axis F1. (See n. 38 on p. 144.)

The solar-time pipe C5 (Fig. 4.17) at the S polar pivot carries a 57-toothed wheel C6, arranged to drive a 59-toothed wheel E3 on an inner lunar-motion pipe E2 by means of an idler-pinion K1 carried in a bracket K2 on the meridian double ring B2 (Fig. 4.12).

Armillary sidereal component

This component corresponds to the 'Three Arrangers-of-Time Instrument' of the classical Chinese armillary spheres.[34] It comprises the following four rings connected rigidly together, and arranged to rotate as a unit in accordance with sidereal time by means of an ingenious gut-line annual-motion auxiliary drive through the N polar pivot:

(i) The solstitial-colure double ring D1 (Fig. 4.12), diameter about 35 cm, is pivoted at the N pole on an inner sidereal-time pipe D3 rotating freely within the solar-time pipe C3, and at the S pole on a bearing rotating freely around the upper end of either the outer solar-time pipe C5 (Fig. 4.17) or the inner lunar-motion pipe E2. An upper quadrant D2 (Fig. 4.12) of one of the twin colure rings D1 is hollow, and conceals part of the gut line H1 by which the annual motion was formerly effected.

(ii) The revolving equator ('red way') single ring D4, diameter also about 35 cm, is engraved on its N face (Fig. 4.15) with the Chinese characters for the 24 *ch'i* or Fortnightly Periods,[35] on its S face (Fig. 4.18) with those for the 28 *hsiu* or Lunar Lodges,[36] and on both faces with $365\frac{1}{4}$ Chinese degrees, one for each day of the year. It retains traces of a red pigment, probably lacquer; whether this is original or was added at some later date has not been determined.

(iii) The ecliptic ('yellow way') single ring D5 (Fig. 4.12), diameter about 36 cm, intersects the solstitial-colure double ring D1 at right angles and the revolving equator ring D4 at an inclination of about $23\frac{1}{2}°$. It retains traces of yellow pigment. It is engraved on its N face (Fig. 4.15) with the Chinese characters for the 24 *ch'i* or Fortnightly Periods, and on its S face (Fig. 4.18) with those for the 28 *hsiu* or Lunar Lodges, as a preliminary to the setting out of which its outer edge appears to have been lightly marked with $365\frac{1}{4}$ Chinese degrees, one for each day of the year. To this outer edge, on top of the

[34] Needham, SCC III: 339ff and Table 31. [35] *Ibid.* 405, Table 35. [36] *Ibid.* 231–59 and Table 24.

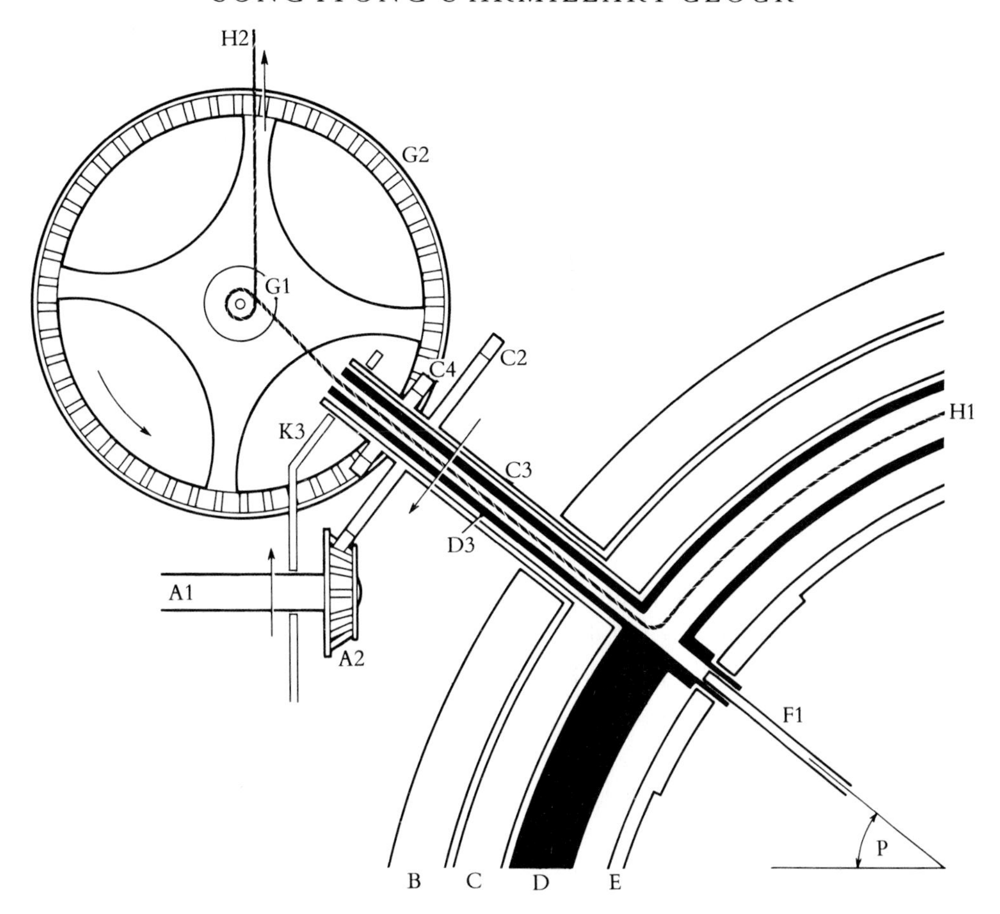

Fig. 4.14. Armillary-sphere north-polar clock-drives: explanatory drawing.

- A1 Main arbor of going train, turning three revolutions a solar day-and-night
- A2 12-leaved bevelled lantern-pinion
- B Fixed terrestrial-coordinate component
- C Solar component, rotated once a solar day-and-night by the clock
- C2 36-toothed wheel
- C3 Outer, solar-time, pipe (see n. 38, p. 144)
- C4 12-leaved pinion
- D Sidereal component, advanced one revolution a year on solar component by action of gut line H1
- D3 Inner, sidereal-motion, pipe (see n. 38)
- E Lunar component, making 57 revolutions in 59 solar days-and-nights
- F1 Polar axis for terrestrial globe (see n. 38)
- G1 Arbor of contrate-wheel, with gut line around its middle section
- G2 48-toothed contrate-wheel, turning one revolution in four solar days-and-nights
- H1 Gut line from sun-carriage C7 (Fig. 4.16) via sun-path channel D6 mounted on ecliptic ring D5, and hollow upper quadrant D2 (Fig. 4.12) of solstitial-colure ring D1 of sidereal component
- H2 Gut line H1 continuing to tension-weight H4 (Fig. 4.12) via metal tube H3 hidden in wooden frame upright
- K3 Upper bearing bracket for solar-time pipe C3
- P Polar altitude approximately $37\frac{1}{2}^{\circ}$

Item C is a schematic reconstruction. The other items survive.

degree-marks, is attached a sun-path ring D6 (Fig. 4.16), of hollow circular channel section with a circumferential slot, marked on both edges of the slot with $365\frac{1}{4}$ Chinese degrees independent of those marked on the body of the ecliptic ring itself. The channel accommodates a sliding sun-model carriage C7, from which a square peg C8 projects radially outwards to engage with the missing polar-pivoted sun-transport ring C1 and to support a missing sun-model C9.

The gut line H1 for the annual-motion drive is anchored at one end to the sun-model carriage C7, and within the sun-path channel D6 it passes once round the ecliptic ring before entering the hollow quadrant D2 (Fig. 4.12) of the solstitial-colure double ring D1 on its way to the annual-motion work (Figs. 4.13, 4.14) above the N polar pivot.

(iv) The lunar path ('white way') single ring D7 (Fig. 4.19) is fixed inside the ecliptic ring D5 (Fig. 4.12) about one-eighth of the distance eastwards from the solstitial to the equinoctial points, at an inclination of about 5° to the ecliptic. It retains traces of white pigment. At the inner edge of the ring there is a separate sliding moon-carriage ring E4 (Fig. 4.19). From the sliding ring E4 two round pegs E5 project radially inwards to embrace the polar-pivoted moon-transport ring E1.

On the S face of the fixed moon-path ring D7 27 round pegs D8, some now broken, are riveted parallel to the axis of the ring. These formerly operated a phase-of-moon mechanism which, with the moon-model itself, is now missing from the sliding moon-carriage ring E4.[37]

Armillary annual-motion work

The annual-motion gut line H1 (Figs. 4.13, 4.14) from the ecliptic circle of the sidereal component, after emerging from the inner sidereal-time pipe D3 at the N polar pivot, is wrapped around the slender middle section of an arbor G1 which is

[37] Jeon suggests (STK, p. 70, after Rufus and Lee, 'Marking Time in Korea', p. 256) that the lunar ring is divided by pegs to mark the 28 'mansions' (Lunar Lodges). The 28 Lodges are, however, of various angular extents (SCC III: Table 24) from 1 to 34 Chinese celestial degrees, whereas the photographs show equally spaced pegs of a number which appears to have been originally odd rather than even, and more likely 27 than either 29 or 25 (cf. Rufus, 'Astronomy in Korea', p. 39; he gives the number as 27 while mistakenly supposing that they indicated the Lunar Lodges). Because of the sidereal motion of the moon-path ring, 27 pegs, but no other number below 54, would serve to operate a phase-of-moon mechanism in monthly cycles of $29\frac{1}{2}$ solar days.

Fig. 4.15. Detail of armillary sphere, viewed from the north-west.

The meridian double ring B2 (Fig. 4.12), on the left, is intersected by the fixed equator single ring B3, on the right, which is marked with 360 European degrees grouped in tens and thirties by longer graduations. The terrestrial-horizon single ring B1 is in the background. Inside, from right to left, are the north face of the rotating equator single ring D4 marked with $365\frac{1}{4}$ Chinese degrees and (not visible in the reproduction) the Chinese characters for the 24 *ch'i* or Fortnightly Periods; the north face of the ecliptic single ring D5 marked with $365\frac{1}{4}$ Chinese degrees and the Chinese characters for the 24 *ch'i* or Fortnightly Periods, and with the square peg C8 (Fig. 4.16) of the sun-carriage C7 projecting radially on the right from the sun-path channel D6 (on the outer edge of D5) at a point in the Fortnightly Period Li Hsia (Beginning of Summer) corresponding to an epoch of midnight 7–8 May; the lunar-path single ring D7 (Fig. 4.19); and the solstitial-colure double ring D1 (Fig. 4.12). The moon-transport single ring E1 (Fig. 4.17) can be seen faintly above the terrestrial globe at the bottom. The characters visible on the ecliptic ring D5 (Fig. 4.12) are, from right to left, Ku Yü (Grain Rain), 21 April; Li Hsia (Beginning of Summer), 6 May; Hsiao Man (Lesser Fullness [of Grain]), 22 May; Mang Chung (Grain in Ear), 7 June.

pivoted horizontally above the N polar pivot, and which carries a 48-toothed contrate-wheel G2 revolved once in four days by the 12-leaved pinion C4 on the solar-time pipe C3. The continuation H2 of the gut line then passes through a metal tube H3 (Fig. 4.12) hidden in a curved channel cut in the wooden upright and subsequently filled. An effective friction grip of the gut line on the arbor G1 is provided by tension applied to its free end by a square weight H4 hanging in a vertical chute H5 inside the clock case near the N polar pivot.

The diameter of the middle part of the horizontal arbor G1 is such that its rotation pulled the gut line H1 out from the N polar pipe, through a length equal to that of one Chinese degree on the ecliptic ring each day. This caused the sidereal component to advance gradually on the solar component so that it actually revolved in accordance with sidereal time, performing in the course of a year one complete rotation more than the solar component. The gut line needed to be rewound onto the ecliptic circle annually.

Now that the polar-pivoted sun-transport ring C1 (Fig. 4.16) is missing, the annual-motion work is ineffective. It has not been possible to ascertain whether the sidereal component is at present free to rotate independently of the solar-time pipes C3, C5 (Figs. 4.14, 4.17) at the N and S polar pivots, or whether it has now been fixed to these vestiges of the solar component.[38]

Armillary lunar component

The lunar component comprises a polar-pivoted moon-transport single ring E1 (Fig. 4.17). This ring is fixed at the S pole to the inner lunar-motion pipe E2 turning between the outer solar-time pipe C5 and the fixed terrestrial axis F1, and at the N pole to a bearing free to rotate around the lower end of the inner sidereal-time pipe D3 (Fig. 4.14). The ring E1 (Fig. 4.17) was rotated by the 59-toothed wheel C6 on the S pole inner lunar-motion pipe E2, coupled to the 57-toothed wheel C6 on the S pole outer solar-time pipe C5 by the idler-pinion K1. Its motion relative to other components of the armillary sphere may be described as follows:

[38] A conversion (or attempted conversion) of the instrument to represent a rotating-earth system could have led to the separation of a globe-carrying upper part of the polar axis from its fixed lower part, and to the making of solid connections between the upper part of the axis and the sidereal- and solar-time pipes at the N polar pivot (see nn. 15 and 31 above).

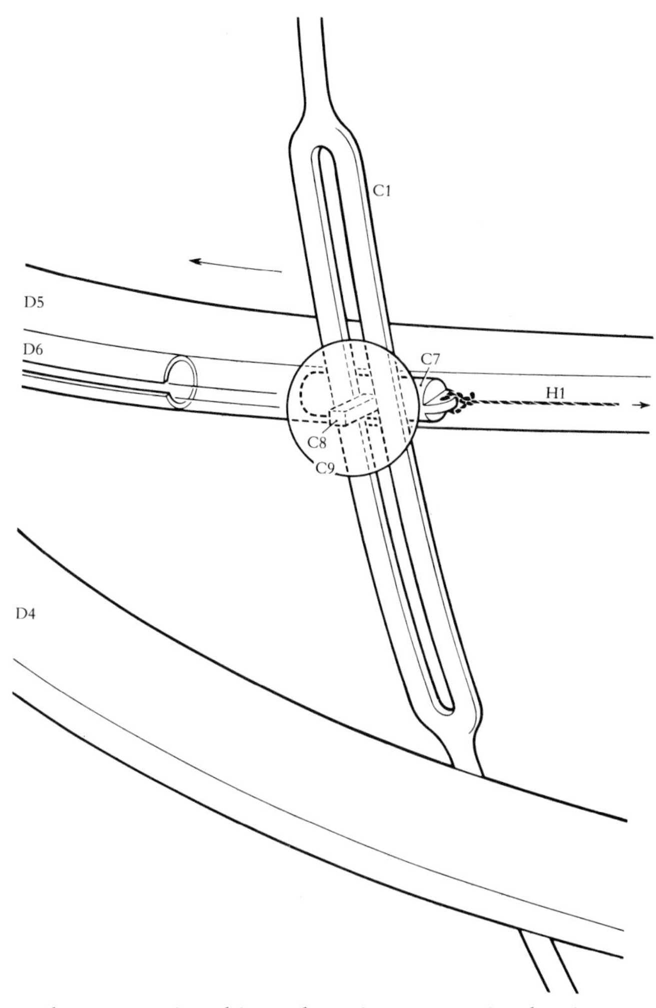

Fig. 4.16. Armillary-sphere sun-carriage drives: schematic reconstruction drawing.

C1 Sun-transport ring, driven leftward (as seen in this view, looking towards centre of armillary sphere near spring equinox) at solar rate by the clock

C7 Sun-carriage, sliding in sun-path channel D6 mounted on ecliptic ring D5 of sidereal component

C8 Square supporting peg for sun-image C9

C9 Sun-image

D4 Equator ring of sidereal component, driven leftward (in this view) at sidereal rate by addition of gut-line annual-motion rate to solar rate of sun-transport ring C1

D5 Ecliptic ring

D6 Sun-path channel, mounted on outer edge of D5

H1 Gut line from sun-carriage C7 to annual-motion work (Figs. 4.13, 4.14) via hollow upper quadrant D2 of solstitial-colure double ring D1, and inner sidereal-time pipe D3 (Fig. 4.12)

Items C1 and C9 are schematic reconstructions. The other items survive.

(i) $28\frac{1}{2}$ forward rotations, relative to the fixed terrestrial axis, for every $29\frac{1}{2}$ forward rotations of the solar component, that is, in each month of $29\frac{1}{2}$ days.
(ii) One backward rotation, relative to the solar component, in each month of $29\frac{1}{2}$ days.
(iii) About $12\frac{1}{2}$ backward rotations, relative to the solar component, in each year.
(iv) About $13\frac{1}{2}$ backward rotations, relative to the sidereal component, in each year – the extra backward relative rotation being due to the extra annual forward rotation of the sidereal component effected by the annual-motion work.

The function of the moon-transport ring E1 was to drive the sliding moon-carriage ring E4 (Fig. 4.19) around the moon-path ring D7 of the sidereal component by means of the two inward-projecting radial pegs E5, which permitted the necessary changes in lunar declination. In the course of about two years, the moon-carriage ring E4 would have completed 27 circuits of the moon-path ring D7, and in so doing each point on it would have made 27 passages past each of the 27 axially aligned pegs D8[39] on the S face of the ring (Fig. 4.18). It seems probable that by this means a missing phase-of-moon mechanism was caused to operate through the required $12\frac{1}{2}$ monthly cycles of each year. The actual design of the missing phase-of-moon mechanism is unknown, but one possibility, which is illustrated in the schematic reconstruction drawing Fig. 4.19, is a spherical moon-model E8 coloured white on one side and black on the other, rotated within a hemispherical shutter E7 by a 29-toothed gear-wheel E9 driven (via a pinion M1 and idler gear-wheel M2) by a star-wheel M3 engaging with 29 of the 27 pegs D8 in the course of $29\frac{1}{2}$ solar days.

The presence of a phase-of-moon mechanism in a much earlier Chinese armillary clock is known from the following extract from the description, in the *Sung shih*, of Wang Fu's waterwheel-driven clock of 1124 C.E.:

> The sun and moon (are shown) following the ecliptic ... (While the heavens rotate once to the left) the moon makes thirteen degrees and a fraction (to the right). Starting bright in the west, its shape is (seen on the instrument) first like a hook, then only the lower half of it is visible in the west, then at the mid-month it becomes full and round, after that the lower half starts to disappear at the west, then only half can be seen in the east, finally at the end of the month it is quite hidden ... According to the old system the

[39] See n. 37 above.

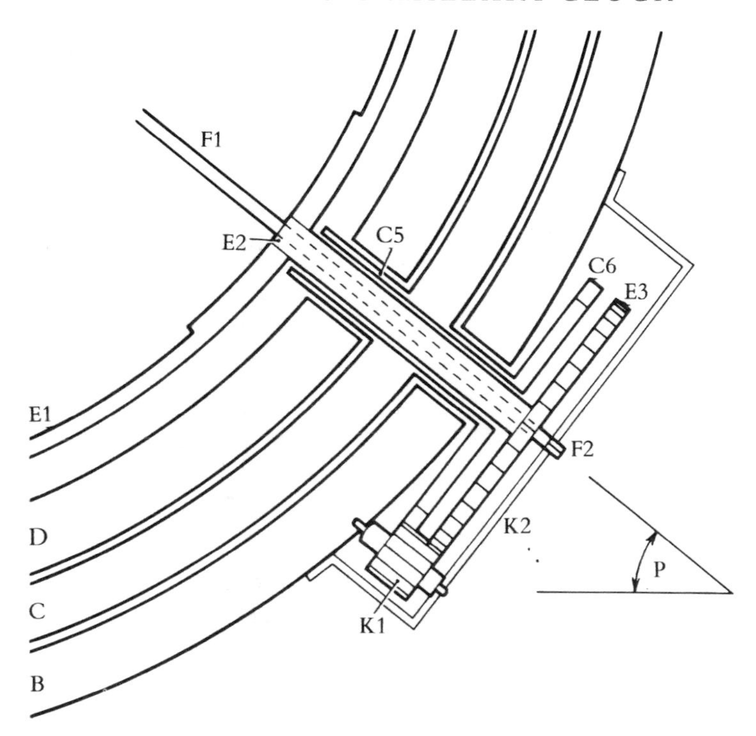

Fig. 4.17. Armillary-sphere south-polar lunar-motion work: explanatory drawing.

B Fixed outer terrestrial-coordinate component

C Solar component, now missing, rotated at solar rate – 59 turns in 59 solar days-and-nights – by the clock

C5 Solar-time pipe

C6 57-toothed solar-time gear-wheel

D Sidereal component, advancing one revolution a year on solar component

E1 Moon-transport single ring, making 57 revolutions in 59 solar days-and-nights

E2 Lunar-motion pipe

E3 59-toothed lunar-motion gear-wheel

F1 Axis for terrestrial globe, Fig. 4.20

F2 Square end, fixing axis F1 to bracket K2 at south pole of terrestrial-coordinate component B

K1 Idler-pinion, coupling C6 and E3

K2 Bracket for idler-pinion pivot-hole, and for fixing square end F2 of terrestrial globe axis F1

P Polar altitude approximately $37\frac{1}{2}°$

Fig. 4.18. Detail of armillary sphere, viewed from the south-south-west.

The rings visible are, from left to right, the fixed equator single ring B3 (Fig. 4.12) engraved with 360 European degrees grouped in tens and thirties by longer graduations; a small part of the revolving equator single ring D4, engraved with $365\frac{1}{4}$ Chinese celestial degrees; the sun-path channel D6 engraved with $365\frac{1}{4}$ Chinese celestial degrees and mounted on the edge of the ecliptic single ring D5; the south face of the ecliptic single ring D5, engraved with $365\frac{1}{4}$ Chinese celestial degrees and with the Chinese characters for the 28 *hsiu* or Lunar Lodges; the lunar-path single ring D7 (Fig. 4.19), with some of the 27 axially aligned pegs D8 for operating the missing phase-of-moon mechanism, Fig. 4.19; the sliding moon-carriage ring E4; and the polar-pivoted moon-transport single ring E1. The characters visible on the ecliptic single ring D5 (Fig. 4.12) are, from bottom to top, those for *hsiu* numbers 9, (*ch'ien*) *niu*; 10, *nü*; and 11, *hsü*. The last of these is engraved in a non-standard form: 虗 in place of the usual 虛.

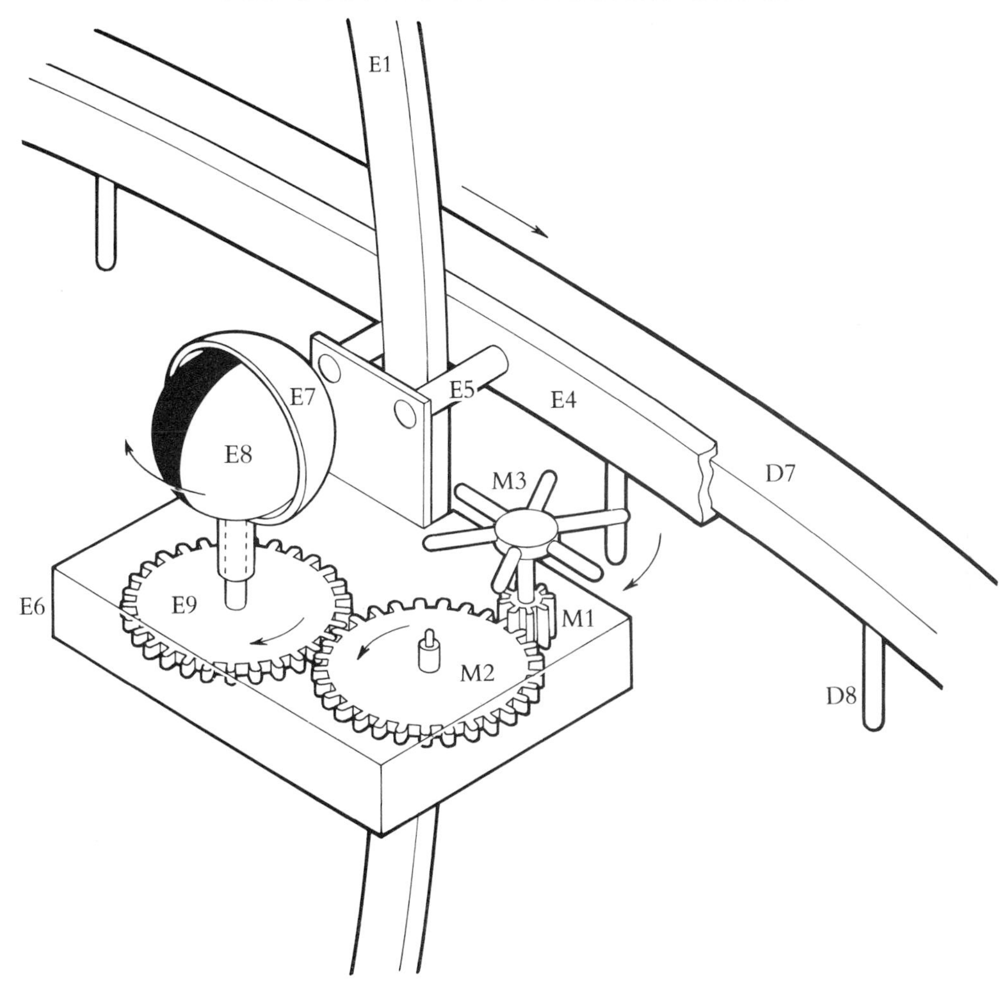

Fig. 4.19. Armillary-sphere phase-of-moon mechanism: schematic reconstruction drawing.

- D7 Moon-path single ring, carried by sidereal component at sidereal-time rate of $366\frac{1}{4}$ turns clockwise (as seen from north polar pivot) in $365\frac{1}{4}$ solar days-and nights
- D8 One of 27 axially aligned pegs on south face of D7
- E1 Moon-transport single ring, revolved by lunar-motion work, Fig. 4.17, at lunar rate of $28\frac{1}{2}$ turns clockwise (as seen from north polar pivot) in $29\frac{1}{2}$ solar days-and-nights
- E4 Moon-carriage ring, driven by E1 via pegs E5
- E5 One of two radial pegs on inner face of E4
- E6 Moon-carriage, supported by pegs E5
- E7 Hemispherical shutter for moon-image E8
- E8 Moon-image, drawn as seen at first quarter, rotating on its own axis once in $29\frac{1}{2}$ solar days-and-nights
- E9 29-toothed gear-wheel
- M1 6-leaved pinion
- M2 Idler gear-wheel
- M3 6-armed star-wheel. The relative motion between moon-path ring D7 and moon-carriage ring E4 causes this star-wheel to engage $27 + 2 = 29$ of the 27 axially aligned pegs D8 every $29\frac{1}{2}$ days-and-nights, so rotating the moon-image on its axis to show the phases correctly

Items D7 to E5 survive. Items E6 to M3 are schematic reconstructions.

body of the moon was always round, so that one could not distinguish the phases. Now it is turned by the mechanism in such a way that it sometimes looks round, sometimes crescent-shaped; sometimes dark and sometimes visible; all in agreement with the phenomena of the heavens.[40]

The terrestrial globe (Figs. 4.20–1)

The globe is of wood, with an oil-painted paper surface. Its diameter is approximately 9 cm. Besides geographical markings, names in Chinese characters, and certain early voyages of discovery, it has meridians and parallels spaced at intervals of ten European degrees. The geographical information could have been obtained from a map of the world which was presented by the Jesuit missionary J. A. Schall von Bell to the Korean crown prince during the latter's residence in Peking following the war of 1644.[41] The globe was designed to be held stationary, presumably with Korea horizontal at the top, by the engagement of the square lower end F2 (Fig. 4.17) of its supporting axis F1 with a square hole in the bracket K2 holding the lunar-motion work on the outside of the prime-meridian double ring. The N end F1 (Fig. 4.14) of the axis is located in, but was not originally intended to be revolved by, the lower end of the sidereal-time pipe D3 at the N polar pivot.[42]

SUMMARY

While this instrument is strikingly innovative in its incorporation of Western-style clockwork, it is at the same time remarkably faithful to ancient East Asian traditions of horological instrumentation. We have seen that the middle nest of rings in the armillary sphere is made to rotate, just as it was in the Chinese instruments of the T'ang and Sung, by means of a shaft in the polar axis. The sphere has a horizon ring fixed externally, as did all Chinese armillaries after the time of Chang Heng. It has, as well, an earth-model placed centrally, as had the instruments of Wang Fan and Ko Heng in the third century C.E.; now the earth-model is a globe marked with the chief continents, like that brought to Peking

[40] Needham, Wang, and Price, HC, pp. 119–22.

[41] Rufus, 'Astronomy in Korea', p. 37; Rufus and Lee, 'Marking Time in Korea', p. 255, Fig. 3; Jeon, STK, pp. 301–5, Fig. 5.6.

[42] See nn. 15, 31, and 38 above.

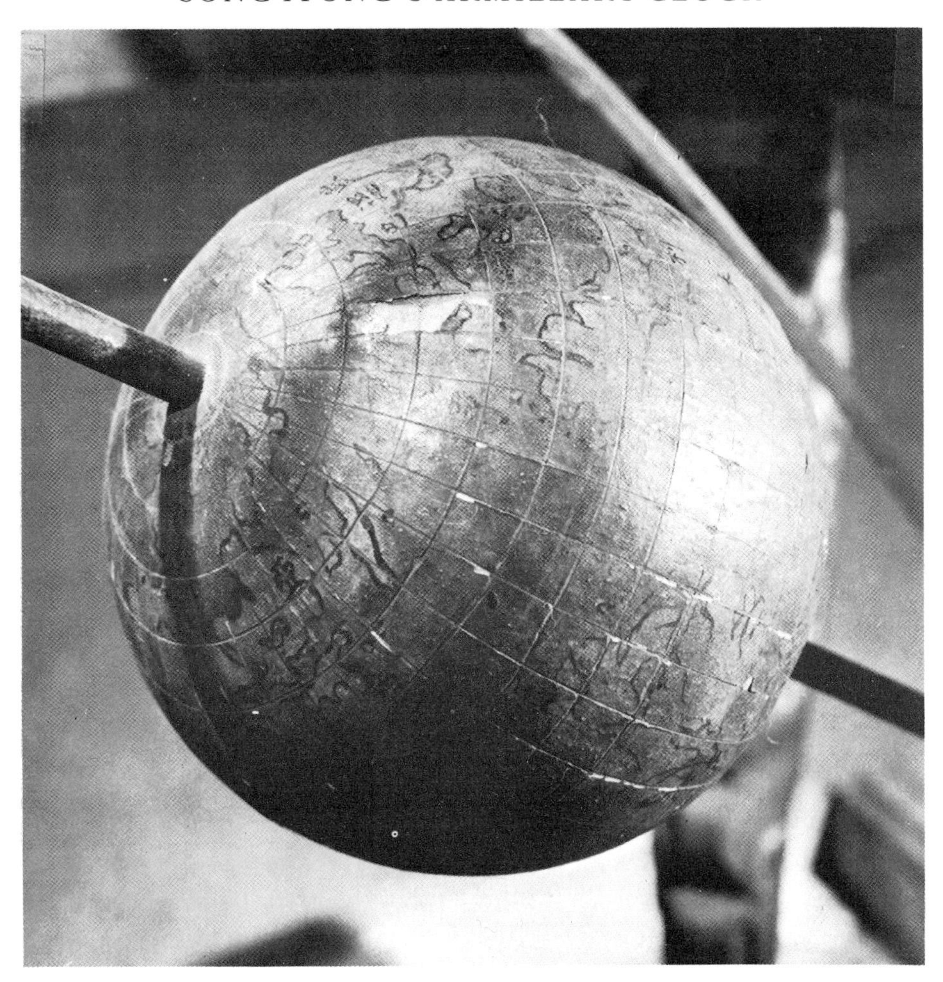

Fig. 4.20. Detail of terrestrial globe, viewed from the north-west.

In this view of the globe the focus is on the North Atlantic (T'ai Hsi-yang). Towards the east is Europe (Ou-la-pa), but the British Isles are missing owing to local damage to the surface of the globe. Towards the west is North America; the cartographer had a very hazy idea of what it looked like. South America was more clearly known; the Amazon is shown with admirable accuracy. The diameter of the globe is approximately 9 cm.

from Persia by Jamāl al-Din in 1267, for the consideration of Kuo Shou-ching. We even find a special ring for the path of the moon, like that which was incorporated in Li Shun-feng's design of 633, as well as in others afterwards. Within the casing, the striking mechanism is operated by the periodic release of balls, as it was in the Striking Clepsydra of King Sejong's time; use of balls goes back to Wang Fu's clock of 1120, and to Arabic water-clocks such as those of al-

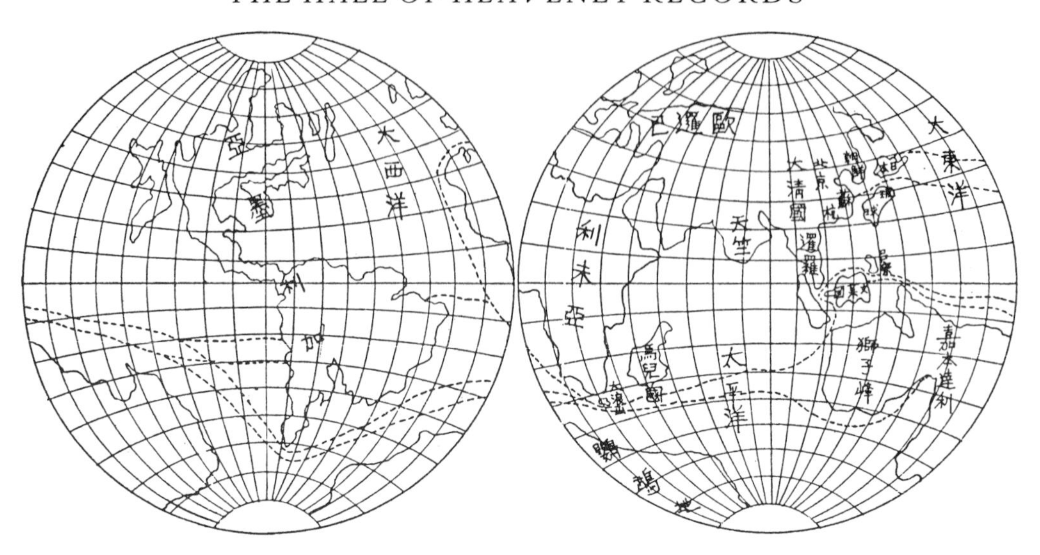

Fig. 4.21. World map of terrestrial globe, redrawn as two planispheres.

This drawing, from Rufus and Lee, 'Marking Time in Korea', p. 257, Fig. 3, shows the world map of the terrestrial globe of Yi Minch'ŏl's armillary sphere with greater clarity than the photograph in Fig. 4.20, but some details – e.g. the Amazon River – are missing here.

Jazarī of the early thirteenth century, or still further to the seventh-century 'steelyard' clepsydra at Antioch, described in the Old History of the T'ang Dynasty (*Chiu T'ang shu* 售唐書) in the tenth century. There is also a noria-like device to raise the balls up again to their reservoir. Finally, there is a window at which discs, the descendants of the traditional time-reporting jacks of the great Chinese clocktowers and Sino-Korean time-announcing clepsydras, present themselves in turn. The Song Iyŏng / Yi Minch'ŏl clock deserves widespread recognition as a landmark in the history of East Asian horology.[43]

[43] This paragraph has been adapted from Needham, SCC IV.2: 520. The 'steelyard' clepsydra is the subject of further discussion in Section 2.2 of the Supplement to Needham, Wang, and Price, HC, 2nd edn, 1986.

5

A KOREAN ASTRONOMICAL SCREEN OF THE EIGHTEENTH CENTURY[1]

In Chapter 1 we mentioned that the founder of the Yi Dynasty, King T'aejo, ordered a planisphere to be engraved in stone and set up in the Bureau of Astronomy; it was completed in 1395 C.E. That planisphere, with its inscription, was one of the fundamental documents of Korean astronomy for the first half of the Yi period; it summarised basic principles of Chinese astronomy on which the Korean science was built.

We have found no records of any new planispheres being made in Korea, either as a part of King Sejong's re-equipping of the Royal Observatory or thereafter, until the late seventeenth century, although we may be sure that rubbings of the 1395 stele were taken from time to time and other copies produced from those by hand. Having been silent on the subject of planispheres hitherto, we turn to it at this point, to describe in detail an eighteenth-century 8-panel folding screen that bears not only a copy of King T'aejo's 1395 planisphere but a double planisphere of Jesuit origin as well. This screen (Fig. 5.1) is a striking embodiment of the way in which traditional Chinese astronomy began to fuse in the eighteenth century with the astronomy of Renaissance Europe to form one modern science.

The first three panels on the right are occupied by a beautiful reproduction of the planisphere of 1395; the next four panels are taken up by two planispheres, one for the northern and one for the southern hemisphere, prepared in the Jesuit period. These are accompanied by two long inscriptions, one above and one below. The last panel to the left is occupied by a series of mid-eighteenth-century diagrams of the sun, moon, and principal planets.

[1] An earlier version of this chapter appeared as: Joseph Needham and Lu Gwei-djen, 'A Korean Astronomical Screen of the Mid-Eighteenth Century from the Royal Palace of the Yi Dynasty (Chosŏn Kingdom, 1392 to 1910)', *Physis*, 1966, *8.2*: 137–62.

THE FOURTEENTH-CENTURY PLANISPHERE

The title is at the top on the right: 'Ch'ŏnsang yŏlch'a punya to' ('T'ien-hsiang lieh-tz'u fen-yeh t'u' 天象列次分野圖). It may be translated as 'Positions of the Heavenly Bodies in their Natural Order, and their Allocated [Celestial] Fields'. The planisphere is shown in Fig. 5.2.

The stars, painted in colour, are arranged in the classical Chinese constellations.[2] Some attempt is made to represent different magnitudes, and constellations are rendered in the usual way by connecting the component stars with thin lines. The colours of the constellations vary, their stars being painted either red, blue, or yellow (perhaps gold), in accordance with their traditional attributions to the three great fourth-century B.C.E. astronomers Shih Shen 石申, Kan Te 甘德, and Wu Hsien 巫咸.[3] The projection is equatorial-polar, the equator being represented by a complete red circle within the planisphere.[4] The ecliptic is shown as a complete yellow circle intersecting with it. The north polar region ('Purple Tenuity Palace', *tzu wei kung* 紫微宮; see Chapter 1, p. 4 above) is demarcated by a circle at about 38° north polar distance. The *hsiu* 宿 (Lunar Lodges) are represented by truncated sectors of unequal breadth. The determining *hsiu* constellations will always be found in the neighbourhood of the equator. The course of the Milky Way is clearly shown, but not in such dark colour as on some versions of this planisphere.

The peripheral band, which could represent a girdle round the heavenly vault in the southern hemisphere, or a circle of perpetual invisibility towards the south celestial pole at about −55° decl., is divided into twelve 'palaces' (*kung* 宮). Each of these is labelled in three ways, first by the Chinese names for the twelve divisions of the Greek zodiac, secondly by using the twelve Chinese cyclical signs (*chih* 支), and thirdly by the names of a series of ancient Chinese proto-feudal states, which according to traditional astrology were governed by the stars in those quarters.[5] For example, segment 1 reads *Pai yang kung*, *hsü*, *Lu chih fen* 白羊宮戌魯之分, meaning 'The Palace of Aries, the cyclical sign *hsü*, the allocation

[2] Complete tabulated identifications will be found in W. Carl Rufus and Celia Chao, 'A Korean Star Map', *Isis*, 1944, *35*: 316–26.

[3] See Needham, SCC III: 263. The practice began in the middle of the fifth century C.E.

[4] The full circle is divided into $365\frac{1}{4}^{d}$.

[5] Needham, SCC III, 545. Note that the term *kung* is used here for the Jupiter Stations rather than for the more ancient division of the sky into five segments to which that term is usually applied.

Fig. 5.1. General view of the Korean astronomical screen, photographed at Seoul.

[corresponding to the State of] Lu'. Or again, segment 6 reads *Shuang nü kung, ssu, Ch'u chih fen* 雙女宮巳楚之分, i.e. 'The Palace of Virgo, the cyclical sign *ssu*, the allocation [corresponding to the State of] Ch'u'.

The use of the Chinese names for the Greek zodiacal signs is extremely rare in documents of the classical Chinese astronomical tradition. They became known in East Asia only in the eighth century, from Indian Buddhist texts.[6] Here

[6] Nakayama, *A History of Japanese Astronomy*, pp. 59–62. The standard Chinese terms as found in modern reference works differ only slightly from those on the screen. See also W. Carl Rufus, 'The Celestial Planisphere of King Yi Tai-Jo', *Transactions of the Korea Branch of the Royal Asiatic Society*, 1913, *4*: 23–72 (abridged version published as 'Korea's Cherished Astronomical Chart', *Popular Astronomy*, 1915, *23.4*: 193–8).

they were certainly introduced in the 1395 revision of the Korean planisphere (see below); and they were inappropriate to the astronomy of that planisphere, because everything in it is equatorial, not ecliptic.[7]

The inscription at the bottom on the left reads as follows:

> What this Bureau of Astronomy [the Sŏun Kwan 書雲觀] has preserved is the reproduction (of the star-chart) that was engraved in stone (in ancient times). (In 1395) a copy [rubbing] of the old star-chart at Kisong 箕城 [Pyŏngyang] was presented to our king T'aejo. The king treasured it greatly and ordered the Bureau to have it (re-)engraved in stone.

This inscription can be understood only in the light of a good deal of further information.[8] According to an old tradition, for some centuries a planisphere engraved in stone had been preserved at Pyŏngyang, but it was lost in the river in 672 during the disturbances when the State of Koguryŏ fell to the State of Silla.[9] When the Yi Dynasty was founded in 1392, the first ruler, T'aejo (r. 1392–8), did all he could to restore the arts and sciences. He was very pleased therefore when one of his subjects in whose family a good rubbing of the original stone had been preserved presented it to the court. The king then ordered the Bureau of Astronomy to have it re-engraved, and this was done with the addition of suitable modifications to bring it up to date. The resulting stele, dated 1395, still exists; it is rather severely damaged.[10] The *Chŭngbo munhŏn pigo* says that this stone was first kept at the Kyŏngbok Palace 景福宮.[11] In 1434 under King Sejong, a special gallery, the Hŭmgyŏnggak 欽敬閣 (see Chapters 2 and 3 above) was built near the Kangnyong Hall 康寧殿 of the Palace, and the stele was placed therein.[12]

[7] They do not appear, for example, on the great Suchou planisphere of 1193 engraved on stone in 1247; this is otherwise very comparable, even in orientation, with the Korean one of 1395. See W. Carl Rufus and Hsing-chih Tien, *The Soochow Astronomical Chart* (Ann Arbor: University of Michigan Press, 1945), and more briefly Needham, SCC III: 278, Fig. 106. The most thorough study is by P'an Nai 潘鼐, 'Suchou Nan-Sung t'ien-wen t'u-p'ai ti k'ao-shih yü p'i-p'an' 苏州南宋天文图碑的考释与批判 (Study and critique of the Southern Sung planisphere stele at Suchou), *K'ao-ku hsüeh-pao* 考古学报, 1976.1: 47–61.

[8] See Rufus, 'Celestial Planisphere', pp. 27ff; Jeon, STK, pp. 22–4; also Ch. 1 above.

[9] References in Rufus, 'Celestial Planisphere', p. 37. This planisphere and its accompanying inscriptions may well have originated in Sui or T'ang China, for we have detailed records of similar scientific transmissions during the seventh century. See Rufus, 'Astronomy in Korea', pp. 12, 14. Recent Chinese studies of historical planispheres possibly related to the Korean one of 1395 are cited in Xi Zezong, 'Chinese Studies in the History of Astronomy, 1949–1979', pp. 464–5.

[10] Jeon, STK, pp. 26–8.

[11] CMP, 3: 29.

[12] Rufus, 'Astronomy in Korea', p. 31. This building also housed the Jade Clepsydra; see Ch. 2 above.

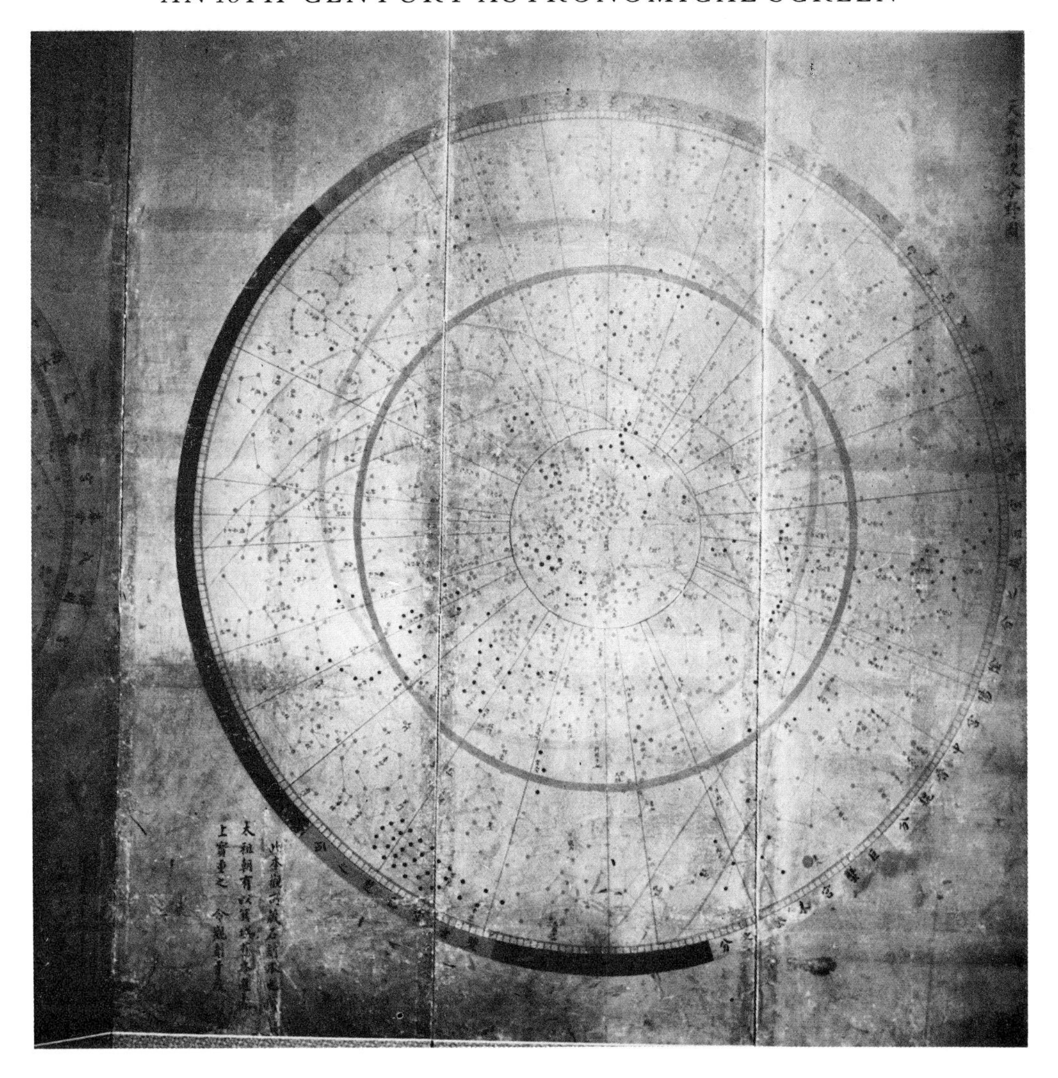

Fig. 5.2. The Korean planisphere of 1395, occupying the three right-hand panels of the screen.

Afterwards this structure was destroyed by fire, rebuilt, and then again destroyed in 1592 at the time of the Japanese invasion. Under King Sukchong (r. 1674–1720) a new stone stele was cut from pure white marble, 7 feet high and 3 feet wide, to perpetuate the data on the old one.[13] This was probably done *c.* 1687, when Yi Minch'ŏl was repairing armillary clocks for the king. Nearly a hundred years later, in 1770, King Yŏngjo (r. 1724–76) had the old stone stele moved to be with

[13] Rufus, 'Astronomy in Korea', p. 42; Jeon, STK, p. 28.

the new one in a new Hŭmgyŏnggak gallery of the Ch'angdŏk Palace (昌德宮, which replaced the ruined Kyŏngbok Palace in the early seventeenth century), and the history of both was recorded on a wooden tablet, since lost. They were studied at the Ch'angdŏk Palace in 1913 by W. Carl Rufus, to whom we owe much of our knowledge about them.[14] Both steles are dated 1395, and the younger monument is a faithful copy of the older one.

The making of the star-chart of 1395 is recorded upon it as follows:[15]

A version of the astronomical chart above engraved in stone was anciently kept at Pyŏngyang, but on account of the disturbances of war it was sunk in the river and lost. Many years passed by and even rubbings of it were no more to be found.

However, at the beginning of the present dynasty a man who had one of these originals presented it to the throne. His Majesty prized it very highly and ordered the Bureau of Astronomy to engrave it anew in stone. The astronomers replied that the chart was very old and that differences had come about in the star-positions (and seasonal correspondences) so that it would need to be modified in accordance with (new) computations (*tui pu* 推步). New determinations of the precise points of solstices and equinoxes would have to be made, and new data collected for dusk and dawn star culminations. Only then could a new star-chart, designed for the future, be recorded and engraved in stone.

His Majesty responded, 'Let it be so!'

It took (several years) until the sixth month of the *i-hai* 乙亥 year (1395) to prepare the new *Chungsŏnggi* (*Chung-hsing chi*) 中星記 (Record of Meridian Transits) and this was then presented to the throne.[16] On the old star-chart at the Fortnightly Period (*ch'i* 氣) 'Beginning of Spring' the Lunar Lodge Mao 昴 (the Pleiades) culminated at dusk; but now Wei 胃 does. As the twenty-four Fortnightly Periods one after the other thus showed (correlation) differences, they used the old chart but corrected the (correspondence with) the (standard) culminating stars. Then the stone was engraved and has just been completed.

Thereupon His Majesty commanded me, his obedient servant and official (Kwŏn) Kŭn 權近, to make a record which should come after the other inscriptions. I, Kŭn, called to mind that from ancient times the emperors did not neglect the veneration of Heaven and its Regulators, and made it their first duty to arrange the calendar, the celestial patterns, and the seasons of labour. Just so emperor Yao commanded Hsi and Ho to set in order the

[14] 'Celestial Planisphere' and 'Astronomy in Korea'; Rufus and Chao, 'A Korean Star Map'.

[15] Tr. Rufus, 'Celestial Planisphere', p. 31, emended from a text on a scroll-painting of the stele preserved at the Old Ashmolean Museum of the History of Science, Oxford. All of the other inscriptions are also translated in this article by Rufus.

[16] This is preserved in *Chŭngbo munhŏn pigo* (CMP), 3: 30ff.

four seasons,[17] and emperor Shun examined the *hsüan-chi* and the *yü-heng*[18] in order to bring into accord the Seven Regulators, faithfully venerating Heaven and diligently serving the people. In my respectful opinion these duties are not to be neglected.

His wise, beneficent, martial, and brilliant Majesty ascended the throne upon the abdication of his predecessor, and his reign has brought peace and prosperity throughout the whole country, with virtue comparable to that of the emperors Yao and Shun. His Majesty placed astronomy in the forefront of his investigations, revising the meridian transit times of stars just as Yao and Shun did for their Regulators. In this way, I believe, by observing the heavenly bodies and constructing astronomical instruments, His Majesty sought to find out the mind of Yao and Shun and to emulate their most worthy example.

His Majesty demonstrated this pattern to the hearts of all; upwards by observing the heavens and the seasons, downwards by diligently serving the people. Thus through His Majesty's spiritual achievements and prosperous zeal he stands together with the two emperors highly exalted. Moreover, he has caused this star-chart to be engraved in fine stone to be a perpetual pleasure for his sons and grandsons for ten thousand generations.

May all who read accept this! Kwŏn Kŭn received the royal ordinance to make this record, Yu Pangt'aek 柳方澤 to supervise the computations and Sŏl Kyŏngsu 偰慶壽 to write the inscriptions.[19] The names of the members of the Bureau of Astronomy (who collaborated in the work) follow: Kwŏn Chunghwa 權仲和, Ch'oe Yung 崔融, No Ŭlchun 盧乙俊, Yun Inyong 尹仁龍, Chi Sinwŏn 池臣源, Kim T'oe 金堆, Chŏn Yungwŏn 田潤權, Kim Chasu 金自綏, and Kim Hu 金侯.[20]

Dated in the 28th year of the Hung-wu 洪武 reign-period [of the Ming dynasty of China], in the twelfth month [December 1395].

THE JESUIT PLANISPHERES

The title of these is 'Huang-tao nan-pei liang tsung hsing-t'u' 黃道南北兩總星圖, i.e. 'General Map of the Stars in both Northern and Southern Hemispheres on Ecliptic Coordinates'. The first striking thing about these planispheres (see Figs. 5.3–5) is that the constellations are all the traditional Chinese ones and not the European, with the exception of a number of asterisms near the south celestial

[17] Needham, SCC III: 186ff.

[18] The identification of these instruments is discussed in Ch. 2, n. 11.

[19] Sŏl's elder brother Chang-su 長壽 had been the writer of the *Chungsŏnggi* and the provisional planisphere which accompanied it to the throne (Rufus, 'Astronomy in Korea', p. 23). He himself was a renowned calligrapher.

[20] The elaborate titles of all these officials are here eliminated.

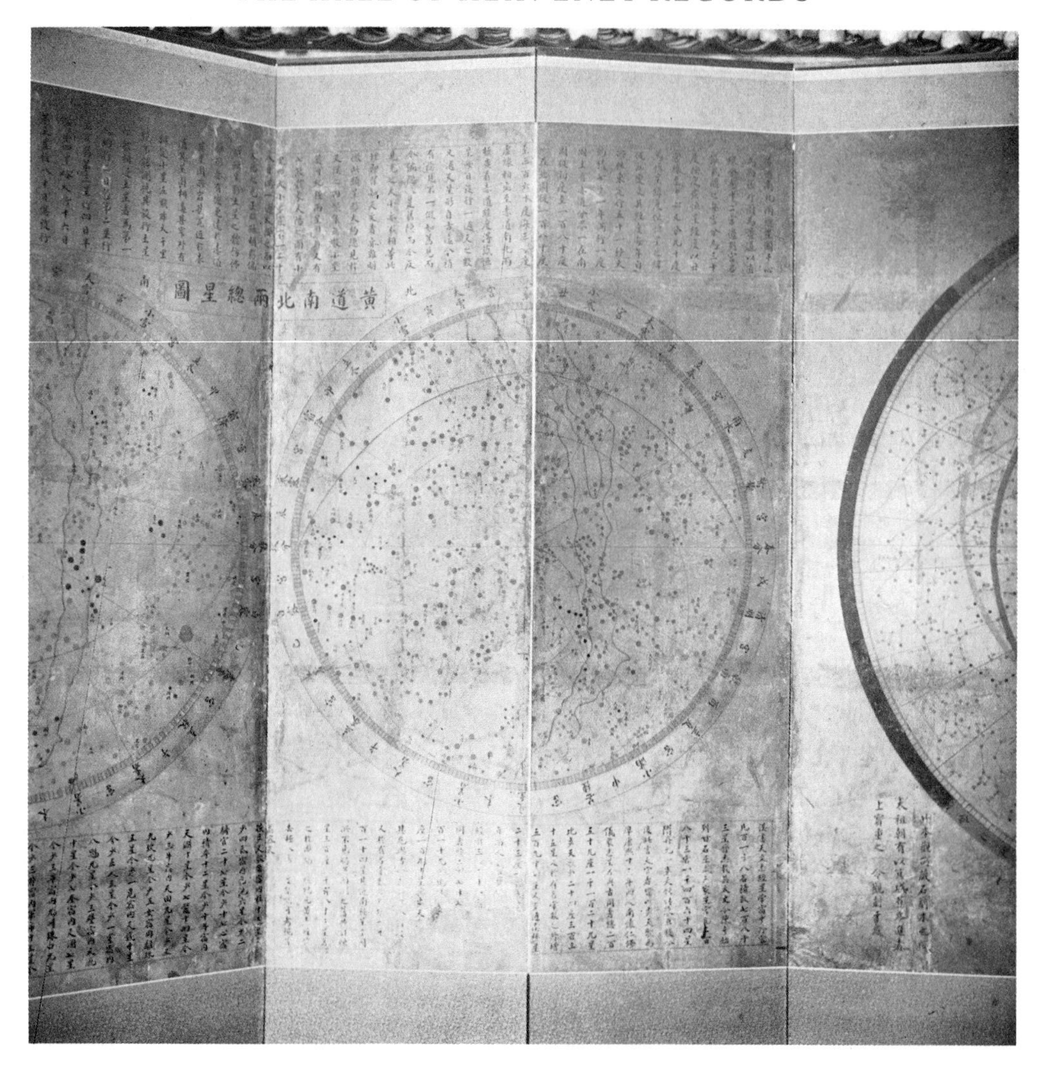

Fig. 5.3. The central panels of the screen, showing the upper and lower Jesuit inscriptions, the northern planisphere, and a portion of the southern planisphere.

pole which had not been recorded by Chinese observers. And not only this, but the three colours attributed to the ancient astronomers are also retained – a highly conservative proceeding (see p. 154 above). The constellations are very clearly seen in Fig. 5.4, which shows the Milky Way running vertically across the planisphere at its nearest point to the south ecliptic pole. The coordinates are ecliptic, i.e. those of celestial longitude and latitude, but the equator is also marked as a graduated red line, while there are lighter blue circles for the tropics and for the declination circles. It is interesting too that on the peripheral band round the ecliptic at the edge of both planispheres zodiacal signs are not given (though they would have been quite appropriate); instead there are the twenty-four Fortnightly Periods and the twelve Chinese cyclical signs (*chih* 支, the Earthly Branches), each termed a Palace (*kung*). Thus the eighteenth-century Jesuits, who, as we shall see, were mainly responsible for the making of these planispheres, imported the Greek (not the modern) coordinate system. On the other hand, they were also faithful to elements of the Chinese tradition. They retained all its constellation patterns and marked the ecliptic in an entirely Chinese way, though in its graduation they adhered to European practice in that 360° was taken for the circle rather than the $365\frac{1}{4}^{d}$ of Chinese tradition.

The upper inscription

'General Map of the Stars in the Northern and Southern Hemispheres on Ecliptic Coordinates'.

These general star-charts have as their centre the two poles of which the ecliptic is the circumference. Radial lines separate the twelve palaces [zodiacal regions]. Their names are entered in the borders and those of the (24) Fortnightly Periods added conformably. Each zodiacal region (naturally) occupies 30°. If you want to ascertain the degree of celestial latitude of any fixed star you measure along the longitude line of (the cyclical sign) *ch'ou* 丑, which from centre to circumference is divided into 90°; this gives the latitude. The latitude of the fixed stars never changes, but each year their longitude moves from west to east by 51″ approximately, making in seventy-one years one degree. On the chart the dividing line of the equator is marked; in the southern hemisphere it runs from (longitude) 0° to 180° and in the northern from 180° to 360°. Every 30° a lighter line crosses it joining it to the equatorial poles north and south, from which can be obtained the right ascension. One can see by this how the fixed stars make one (apparent) complete revolution in one day (and night).

The forms in the heavens also have their cycles and changes; from ancient times until now there have been certain changes in visibility. For example, some that were seen of old are now partly hidden, and others that were not before seen are now quite brilliant, so that the sizes seem not to have remained the same. Even learned astronomers find it difficult to understand the reasons for this. (Variable) stars of this sort are generally to be seen in the Milky Way, where we find collected together countless numbers of small stars.

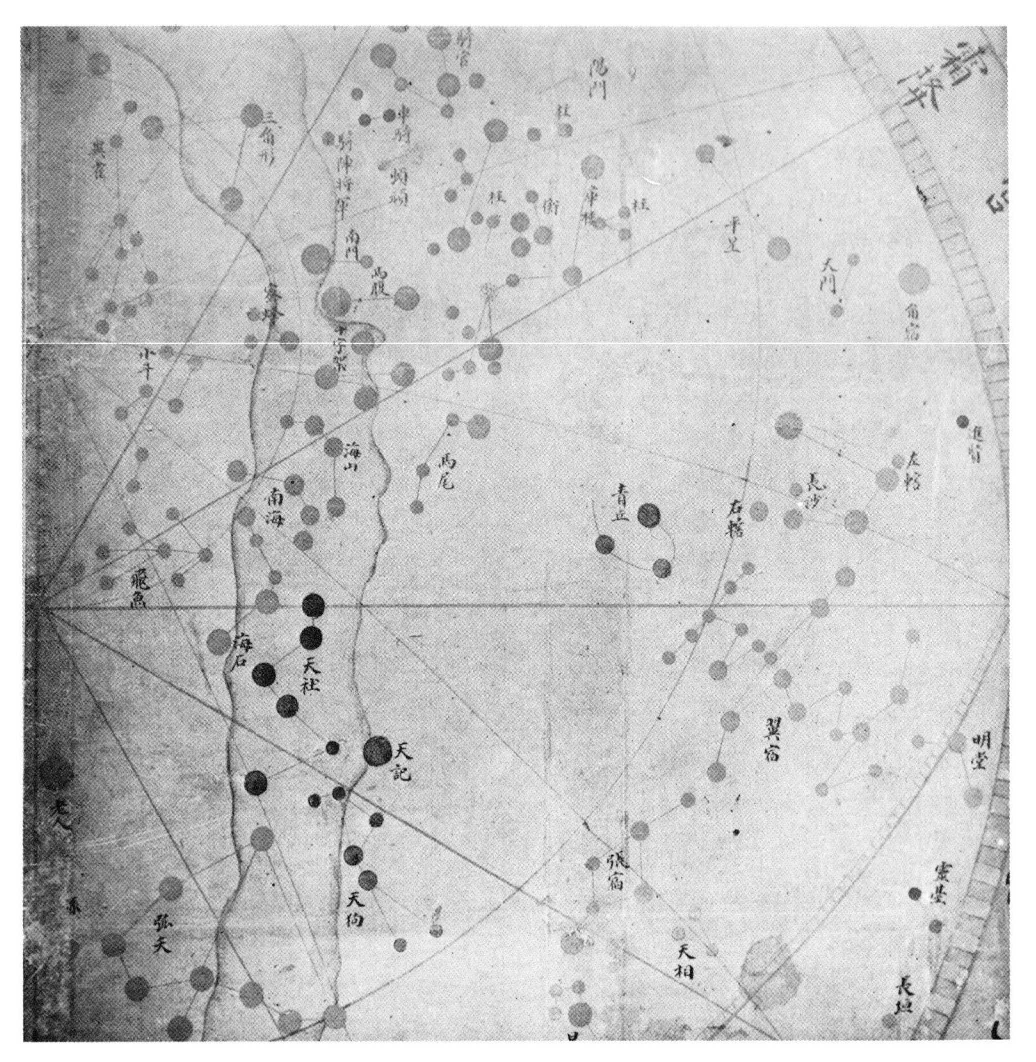

Fig. 5.4. Detail of the southern planisphere.

For orientation, Canopus (= 'Greybeard', Lao-jen 老人 (Schlegel, no. 205)) is just in the picture at the lower left, and the Southern Cross (Schlegel, no. 36) is clearly seen well above and to the left of the centre of the photograph. Above it there are the constellations 'Southern Gate', Nan-men 南門 (Schlegel, no. 244), and 'Horsebelly', Ma-fu 馬腹 (Schlegel, no. 228). Just below it are 'Ocean Mountain', Hai-shan 海山 (Schlegel, no. 78), and 'Horsetail', Ma-wei 馬尾 (Schlegel, no. 230). All these are painted in red, but further below come three constellations all painted in dark blue, 'Altar of Heaven', T'ien-she 天社 (Schlegel, no. 434), 'Heavenly Recorder', T'ien-chi 天記 (Schlegel, no. 470), and 'Heavenly Dog', T'ien-kou 天狗 (Schlegel, no. 459).

Besides (the fixed stars shown in these) two star-charts there are the 'Seven Regulators' [i.e. the sun, moon, and five visible planets].[21] As regards the form of the sun, on its surface there are black spots, large and small, never the same in number, and with a cycle of 28 days. The face of the moon reflects the light of the sun; when it falls directly on it, it is at the full, when it shines upon it sideways there is a black shadow. The body of Saturn is shaped somewhat like an egg, having slight changes in form. When it is furthest away from the equator the circumferential part is at its broadest, like the gallery around a house; when it is very near the circumferential part looks the most narrow. Attached to Saturn there is a series of five small moons [lit. stars], the rotation of which you cannot observe without using a large telescope. The first one, that nearest to it, has a period of revolution of $1\frac{7}{8}$ days; the second goes round in $2\frac{7}{8}$ days; the third in $4\frac{5}{8}$ days; the fourth is comparatively larger and takes 16 days; while the fifth takes 80 days. (In these times) each of them circles Saturn once. As for Jupiter, it frequently has a black shadow that moves transversely across its surface. [This must refer to Jupiter's planetary phases.] In addition Jupiter has four small moons [lit. stars]. The first (and nearest) one takes 1 day 73 quarter-hours[22] to revolve, the second 3 days and 53 quarter-hours; the third, which is comparatively larger, takes 7 days and 16 quarter-hours; the fourth takes 16 days and 72 quarter-hours. (In these times) each of them circles Jupiter once. The surface of Mars also has an indefinite black shadow [its planetary phases]. So also Venus and Mercury borrow the light of the sun just as the Moon does in its waxing and waning.

Written in the 1st year of the Yung-cheng 雍正 reign-period [of the Ch'ing dynasty of China, (1723)], the year order being in (the cyclical signs) *kuei-mao* 癸卯, by the Far Westerner Tai Chin-hsien 戴進賢 [Ignatius Kögler, S.J.].[23] Engraved by Li Pai-ming 利白明 [Fernando Bonaventura Moggi, S.J.].[24]

Little comment is required on the foregoing translation of the upper inscription, which gives quite a good account of the planispheres. The opening statement about precession is apposite, since it explains the modifications which the Korean astronomers had been obliged to introduce 328 years earlier into their still older

[21] See Ch. 2, n. 11.

[22] In the old Chinese reckoning each of our (Western) 24-hour periods was divided into 12 double-hours and also into 100 equal intervals called *k'o*; here, however, the divisions are integral (Western) quarter-hours. China and Korea moved to a 96-quarter-hour system with the adoption of the Shih-hsien calendar, which was promulgated in China in 1645 and formally accepted by Korea in 1651. See above, Ch. 1, pp. 8–9, and Ch. 3, p. 103. The correct whole numbers of quarter-hours, in excess of whole numbers of days, for the planetary revolutions are in fact 74, 53, 15, and 66 respectively, not 73, 53, 16, and 72 as in the Jesuit text.

[23] L. Pfister, *Notices biographiques et bibliographiques sur les Jésuites de l'ancienne mission de Chine* (1552 to 1773), 2 vols. (Variétés Sinologiques, nos. 59–60; Shanghai: Mission Press, 1923–4), no. 297. Kögler's literary name was Tai Chia-pin 戴嘉賓.

[24] *Ibid.* no. 313.

star-chart. The second paragraph certainly refers to novae and supernovae,[25] perhaps also variable stars, occulting partners, and comets. None of these phenomena were at all understood in the eighteenth century. There follows a reference to sunspots, which were regarded at that time as a great European discovery.[26] The rest of the statement is devoted to the nature of the planets of the solar system (Fig. 5.5), especially the varying appearance of the rings of Saturn, described in a rather striking metaphor taken from Chinese architecture. It is interesting to note that by this time, the first half of the eighteenth century, a Jesuit is emphasising the 'lunar' phases of the planets – one of the most important proofs of the Copernican model. Finally the text describes the moons of the planets, a splendid discovery due indeed to Galileo and his successors. The satellites of Jupiter had been reported to the Chinese in the earliest Jesuit books on the telescope appearing in their language.[27] The satellites of Saturn were of course not known until the second half of the seventeenth century; the largest one, Titan, was the first to be seen, by Huygens in 1655. The next four were discovered by Cassini as follows: Iapetus in 1671, Rhea in 1672, and Dione and Tethys in 1674.

As we shall see below, the original engraving of these planispheres and the description translated above has now come to light (Fig. 5.6).

[25] Cf. Needham, SCC III: 423ff. See also references to recent studies of these phenomena by Xi Zezong, Bo Shuren, and others in Xi, 'Chinese Studies'.

[26] An account of this had been given in Chinese by Galileo's friend Johann Schreck, S.J. (Teng Yü-han 鄧玉函; Pfister, *Notices*, no. 46), in his *Ts'e-t'ien yüeh-shuo* 測天約說 (Brief description of the measurement of the heavens) in 1628 (see Pasquale M. d'Elia, *Galileo in China*, tr. R. Suter and M. Sciascia (Cambridge, Mass.: Harvard University Press, 1960), p. 40; Needham, SCC III: 447). The only thing he omitted to mention was that the Chinese had been making systematic observations of sunspots since at least 28 B.C.E. (See Needham, SCC III: 435; Yunnan Observatory, 'Wo-kuo li-tai t'ai-yang hei-tzu chi-lu ti cheng-li ho huo-tung chou-ch'i ti t'an-t'ao' 我国历代太阳黑子记录的整理和活动周期的探讨 (The recording of sunspot activity in Chinese history and an investigation of periods of activity), *T'ien-wen hsüeh-pao* 天文学报, 1976, *17*.2: 217–27; and other recent work cited in Xi, 'Chinese Studies'.) These records show the 11-year period, and perhaps also the sun's sidereal rotation period of *c*. 25 days.

[27] Notably thé *T'ien-wen lüeh* 天問略 (Explicatio sphaerae coelestis) by Emanuel Diaz, S.J. (Yang Ma-no 陽瑪諾; Pfister, *Notices*, no. 31), published in 1615. See d'Elia, *Galileo*, pp. 18ff; Needham, SCC III: 444. Also the *Yüan-ching shuo* 遠鏡說 (The far-seeing optick glass) by Johann Adam Schall von Bell, S.J. (T'ang Jo-wang 湯若望; Pfister, *Notices*, no. 49), published in 1626. See d'Elia, *Galileo*, p. 36; Needham, SCC III: 445. Mention of the moons of Jupiter was again made in J. A. Schall von Bell's two tractates of 1634, *Hsin-fa li-yin* 新法曆引 (Introduction of the new calendrical science) and *Hsin-fa piao-yi* 新法表異 (Differences between the old and the new (astronomical and) calendrical systems), but here the rings of Saturn were still interpreted as two small accompanying 'stars' or moons. D'Elia, *Galileo*, p. 45 gives tendentious translations of these two titles as 'European' rather than 'New'; the importance of not doing this will be seen from Needham, SCC III: 447ff.

Fig. 5.5. The three left-hand panels of the screen, showing the southern planisphere and the diagrams of the sun, moon, and planets.

Diagrams of the planets

This will be the most convenient place to refer to the drawings of the sun, moon, and visible planets which occupy the eighth panel from the right of the astronomical screen (Fig. 5.5).

From above downwards and in steadily decreasing diameters (though of course in no way proportional to the precise dimensions which could be represented today) they are shown in the following order:

Sun	T'ai Yang 太陽, in red with a row of red spots shown as a ring south of the equator
Moon	T'ai Yin 太陰, white
Saturn	Chen Hsing 塡星, a whitish brown, with ring and five moons
Jupiter	Sui Hsing 歲星, blue, with four moons
Mars	Ying Hsing 熒星, red
Venus	T'ai Po 太白, white
Mercury	Ch'en Hsing 辰星, dark blue

The lower inscription

This has no title. It runs:

In the Astronomical Chapters of the *Ch'ien Han shu* 前漢書 [History of the Former Han Dynasty (206 B.C.E. to 24 C.E.), written by Pan Ku 班固, *c.* 100 C.E.] it is said that the stars in the (28) *hsiu* [Lunar Lodges] both north and south of the equator were assembled in 118 constellations, containing a total of 783 stars.[28] The Astronomical Chapters[29] of the *Chin shu* 晉書 [History of the Chin Dynasty (265–420), written by Fang Hsüan-ling 房玄齡, 635] say that the Astronomer-Royal of the Wu 吳 Kingdom [222–80], Ch'en Cho 陳卓, first listed and made a chart of the stars described [of old, i.e. in the fourth century B.C.E.] by the three scholars Kan (Te), Shih (Shen), and Wu Hsien, amounting to 283 constellations and 1464 stars.[30] Then afterwards in the *Pu-t'ien ko* 步天歌 [The Song of the Sky-Pacers], written by Tan-yuan-tzu 丹元子 [i.e. Wang Hsi-ming 王希明, *c.* 590],[31] the number of stars was the same as that given by Ch'en Cho. Since those times all who have discussed the constellations have taken the *Pu-t'ien ko* as their standard. Now in the thirteenth year of the K'ang-hsi 康熙 reign-period [1674] the Westerner Nan Huai-jen 南懷仁 [Ferdinand Verbiest, S.J.][32] edited an *I-hsiang chih* 儀象志 [Descriptions of Astronomical Instruments],[33] in which the constellations and stars identical with those known of old amounted to 259 of the former and 1129 of the latter. This was 24 constellations and 335 stars fewer than the number in the *Pu-t'ien ko*. But in addition he added 597 stars. Also there are some near the south celestial pole; of these he added 23

[28] This is the estimate of the Astronomer-Royal Ma Hsü, 馬續 *c.* 130 C.E. (see Needham, SCC III: 265).

[29] Tr. in full by Ho Peng Yoke, *The Astronomical Chapters of the Chin Shu* (Paris and The Hague: Mouton, 1966).

[30] Needham, SCC III: 263, 265.

[31] *Ibid.* 201.

[32] Pfister, *Notices*, no. 124.

[33] On this see Needham, SCC III: 452, and n. 58 (vii) below.

constellations comprising 150 stars ... [*lacuna*] ... years [this must be before 1746] the Westerner Tai Chin-hsien [Ignatius Kögler, S. J.] ... [*lacuna*] ... but in addition he added 1614 stars. So those near the south celestial pole which cannot be seen in China but were measured by Westerners, combined with those formerly known, amounted to 300 constellations with 3083 stars [these are the figures of the Ch'ien-lung star-catalogue completed in 1757] ... And after this was all done the celestial latitudes and longitudes, and the declinations and right ascensions, were completely determined for all these stars much more accurately than ever before.

The next section of the text can conveniently be tabulated thus:

Stars of the older Chinese constellations omitted from this Catalogue

Hsiu (Lunar Lodges) of the *Pu-t'ien ko*	Names of the constellations affected[34]	No. of stars	Decrease in no. of stars in this catalogue
Chüeh 角 (Horn)	(T'ien)-chu 天柱 (567) (Heavenly Pillars)	15	4
Ti 氐 (Root)	K'ang-ch'ih 亢池 (134) (Pool of K'ang) [K'ang, 'Neck', is the preceding Lunar Lodge]	6	2
	Chi 騎 (146) (The Riders)	27	17
Hsin 心 (Heart)	Chi-tsu 積卒 (659) (The Serving-Men)	12	10
Tou 斗 ((Southern) Dipper)	T'ien-yuan 天淵 (624) (Heavenly Abyss)	10	7
	Pieh 鼈 (284) (Turtle)	14	3
Niu 牛 (Ox(-Leader))	T'ien-t'ien 天田 (579) (Heavenly Fields)	9	5
	Chiu-k'an 九坎 (168) (The Nine Canals)	9	5
Nü 女 ((Serving-) Woman)	Li-chu 離珠 (209) (Brilliant Pearls)	5	1
Wei 危 (Ridgepole)	T'ien-ch'ien 天錢 (595) (Heavenly Knife-Money)	10	5
	Jen(-hsing) 人星 (128) (The Man(-Star))	5	1
Shih 室 (Chamber)	Pa-k'uei 八魁 (268) (The Eight Chiefs)	9	3
Pi 壁 ((Eastern) Wall)	T'ien-chiu 天廐 (485) (Heavenly Stables)	10	7
K'uei 奎 (Stride)	T'ien-hun 天溷 (452) (Heavenly Pigsty)	7	3
Pi 畢 (Net)	Chiu-chou Shu-k'ou 九州殊口 (171) (Confusion of Tongues) [lit. 'Different Voices of the Nine Prefectures']	9	3
Ching 井 (Well)	Chün-shih 軍市 (179) (Military Marketplace)	13	7
Hsing 星 ((Seven) Stars)	T'ien-chi 天稷 (590) (Heavenly Millet)	5	5
	T'ien-miao 天廟 (524) (Heavenly Temple)	14	14
Chang 張 (Extension)	Tung-ou 東甌 (640) (Eastern Goblet)	5	5
I 翼 (Wings)	Chün-men 軍門 (180) (Military Gate)	2	2
Chen 軫 (Chariot-Platform)	T'u Ssu-k'ung 土司空 (631) (Minister of Works)	4	4
	Ch'i-fu 器府 (148) (Store of Instruments)	32	32

[34] The number in parentheses following the Chinese name of each constellation is the index number in Gustave Schlegel, *Uranographie chinoise*, 2 vols. (Leiden: E. J. Brill, 1875; repr. Taipei, 1967).

The text continues as follows:

The method is to make a large telescope such that one can use both eyes to look into the heavens; then the number of stars is multiplied several dozen times. The definition is very clear. The *Mao hsiu chuan* 昴宿傳 [Treatise on the *Hsiu* (Lunar Lodge) Mao (the Pleiades)][35] says that what looks like seven stars is really 36 stars.[36] (So also) (the Lunar Lodge) Kuei 鬼 ['Ghost'; in Cancer][37] has the nebula called Chi Shih 積尸 ['Pile of Corpses'][38] which was always said to be a *ch'i* 氣 [vapour] like a white cloud; this again is actually 35 stars. You can count them one after another. The same is true of the Chung Nan 中南 ['South Central'] star in (the Lunar Lodge) (Chien-)Niu 牽牛 ['Ox-Leader'; in Aquarius],[39] and the Tung Yü 東魚 ['Eastern Fish'] star[40] and Fu Yüeh 傅說 ['Invocator'] star[41] in (the Lunar Lodge) Wei 尾 ['Tail'; in Scorpio].[42] So also the southern stars of (the Lunar Lodge) Tsui 觜 ['Turtle-Beak', the head of Orion][43] were considered very small and difficult to make out, but now every one can be distinctly seen.[44] And (lastly) in the Milky Way there are an uncountable number of small stars, so crowded together that they give you the impression of a white river.

This inscription dates the screen, for the Ch'ien-lung star-catalogue alluded to (in our square brackets on p. 167 above) was not published until the 1757 edition of the *I-hsiang k'ao-ch'eng* 儀象考成 (see p. 171 below). The screen would therefore appear to have been painted not long after that date (say *c.* 1760), but since it is quite possible that advance information from Jesuit sources may have reached Korea before the publication date, the screen could be as early as about 1755. What evidence we have of Jesuit–Korean contacts about this time (see pp. 178–9 below) might well suggest the earlier date.

The content of this second inscription is essentially astrographic. After referring to the classical star-catalogues of ancient and early medieval China, it speaks of the Jesuit contributions from 1660 onwards, especially the star-lists of Verbiest and Kögler. What the Jesuits did between then and 1760, as part of their official work

[35] Possibly a reference to part of one of the Jesuit tractates, not yet identified. Or one might translate: 'According to tradition, Mao *hsiu* consists of seven stars, but really it has thirty-six.'

[36] Similar statements had been made already in the *Yüan-ching shuo* (1626) and the two *Hsin-fa* tractates of 1634; cf. d'Elia, *Galileo*, pp. 37, 45.

[37] Schlegel, no. 198.

[38] Schlegel, no. 651. This is the Praesepe nebula or open cluster. Its resolution was mentioned in the *Yüan-ching shuo* and the *Hsin-fa* tractates. Cf. d'Elia, *Galileo*, p. 37, with illustration from the former text.

[39] Schlegel, no. 252. The 'South Central Star' is not listed in Schlegel's catalogue.

[40] *Ibid.* no. 758.

[41] *Ibid.* no. 69.

[42] *Ibid.* no. 714.

[43] *Ibid.* no. 686.

[44] This was mentioned in the *Yüan-ching shuo* already. Cf. d'Elia, *Galileo*, p. 37, with illustration from this text.

in the Chinese Bureau of Astronomy, was to drop those Chinese stars and even whole constellations which were inconveniently faint, and to add stars and constellations of higher magnitude in the southern hemisphere which earlier Chinese astronomers had not succeeded in recording.[45] Why our screen should have itemised the omissions so carefully is not obvious, but it is interesting that they bore on only sixteen out of the total of twenty-eight *hsiu* (Lunar Lodge segments), affecting twenty-two constellations, of which six were entirely abrogated.

The rest of the text emphasises once again the revelations of the telescope regarding nebulae and star clusters. Such resolving powers had been, as we have noted, brought to Chinese knowledge from 1626 onwards. But the outstanding example was of course the galactic plane, the Milky Way itself, and from 1615 onwards Jesuit writers never tired of hymning the uncountable number of small stars which the telescope found in it.[46]

THE BACKGROUND OF JESUIT WORK

Who were the Jesuit astronomers mentioned on the screen? The earliest one is Ferdinand Verbiest (Nan Huai-jen, 1623–88). This eminent mathematician and astronomer was a Belgian; he was the second European to be appointed to the post of Director of the Board of Astronomy and Calendar, from 1673 until his death. He thus succeeded Johann Adam Schall von Bell (T'ang Jo-wang 湯若望, 1592–1666); and he was the one who about 1673 reconstructed the Peking Observatory with entirely new instruments, all made in China. Verbiest was a

[45] It does not follow that they never observed them. In 724 an expedition was sent to the South Seas to chart the stars as far as 20° from the south celestial pole (i.e. −70° decl.; cf. Needham, SCC III: 274, where the text is given in translation), but the records were not preserved. Canopus (Lao-jen 老人 (Greybeard); Schlegel, no. 205), at −52° decl. south, had been familiar since the first century B.C.E., as we know not only from the *Shih chi* 史記 (Historical Records), Ch. 27 (tr. Edouard Chavannes, *Les mémoirs historiques de Se-Ma Ts'ien*, 5 vols. (Paris: Leroux, 1895–1905), vol. 3, p. 353), but also from such apocrypha as the *Ch'un-ch'iu wei yüan-ming pao* 春秋緯元命苞 (p. 55b). The Southern Cross, at −60° decl. south, though not in the classical star-lists, was habitually used by Chinese mariners for navigational purposes (see Needham, SCC IV. 3: 565–6).

[46] This started with Diaz's *T'ien-wen lüeh*, continued in the *Yüan-ching shuo* (1626), and was further emphasised in two tractates with similar titles, the *Heng-hsing li-chih* 恆星曆指 (A guide to the astronomy of the fixed stars) put out by J.A. Schall von Bell in 1630, and the *Wu-wei li-chih* 五緯曆指 (A guide to planetary astronomy) by Giacomo Rho, S.J. (Lo Ya-ku 羅雅谷; Pfister, *Notices*, no. 55) in 1634. Further on this, with translations, see d'Elia, *Galileo*, pp. 17ff, 34, 37, 39, 97, but correct the dates which d'Elia gives for Rho's book to those which are given by Li Nien 李儼, *Chung-suan-shih lun-ts'ung* 中算史論叢 (Collected essays in the history of Chinese mathematics), 2nd ser., 5 vols. (Peking: Science Press, 1954), III: 37.

great friend of the K'ang-hsi emperor (r. 1662–1722) and instructed him deeply, by his wish, in many of the natural sciences.

Eight years before the death of Verbiest there was born in Bavaria the later Jesuit whose name appears most prominently on the Korean screen. He was Ignatius Kögler (Tai Chin-hsien, 1680–1746) and he became the sixth Director of the Board of Astronomy and Calendar, from 1720 until his death. In 1744 he designed and superintended the casting of a large and elaborate equatorial armillary sphere for the Observatory at Peking, where it can still be seen today.[47] In 1713 work had started on a large treatise of astronomy and calendrical science, the *Li-hsiang k'ao-ch'eng* 曆象考成, under the editorship of Ho Kuo-tsung 何國宗 and the distinguished mathematician Mei Ku-ch'eng 梅瑴成.[48] This consisted of three sections, comprising theoretical astronomy, practical technique, and astronomical tables; it embodied many advances over the seventeenth-century Jesuit publications in China, but it still retained the Ptolemaic theory in its Tychonic form.[49] The epoch, however, was changed to 1683 from Tycho Brahe's date of 1628. This work was published in 1723 with an imperial preface (though perhaps the printing was not finished until about 1730) as part of a still wider compendium which included treatises on mathematics and music and was entitled *Lü-li yuan-yuan* 律曆淵源 (Profound Source of Harmonic and Calendrical (Calculations)). The date of 1723 is only coincidental with the date of the upper inscription on our screen, for this seems too condensed to be a likely excerpt from the *Li-hsiang k'ao-ch'eng*, with which indeed at that time Ignatius Kögler was not overtly associated. In 1737 Ho Kuo-tsung memorialised the emperor, asking for a revision and enlargement of the *Li-hsiang k'ao-ch'eng*, and this was approved in the following year. The work was carried out by Ignatius Kögler and his Assistant Director, André Pereira (Hsü Mou-te 徐懋德, 1690–1743). The work was largely based on the new discoveries and formulations of Cassini (Ko Hsi-ni 噶西尼) and Flamsteed (Fu Lan-te 弗蘭德), and new star-charts were added. Elliptical orbits now in part replaced the epicycles, but the system was still geocentric: Tychonic, how-

[47] Needham, SCC III: 452.

[48] Cf. Arthur Hummel, *Eminent Chinese of the Ch'ing Period (1644–1912)*, 2 vols. (Washington: Library of Congress, 1943–4), pp. 285, 569.

[49] Alexander Wylie, *Notes on Chinese Literature* (first published Shanghai and London, 1867; 3rd edn, Peking : Vetch, 1939), p. 89.

ever, not classically Ptolemaic.[50] This revision and enlargement was published as the *Li-hsiang k'ao-ch'eng, hou pien* 曆象考成後編 in 1742.

About this time Kögler prepared an official description of the imperial astronomical instruments at the capital, and this appeared as the *I-hsiang k'ao-ch'eng* in 1744. Kögler was assisted in this by another German Jesuit, Augustin von Hallerstein (Liu Sung-ling 劉松齡, 1703–74).[51] In the same year this group of Jesuits embarked upon a new general determination of star positions which became known as the Ch'ien-lung star-catalogue; this took nearly ten years, being finished in 1752. After the death of Kögler, von Hallerstein was joined by two more Jesuit collaborators, the Bavarian Anton Gogeisl (Pao Yu-kuan 鮑友管, 1701–71) and the Portuguese Felix da Rocha (Fu Tso-lin 傅作霖, 1713–81). The last memoirs on the astronomical equipment used were finished in 1754, the emperor himself wrote the preface in 1756, and the whole work came out in the following year as an enlarged edition of the *I-hsiang k'ao-ch'eng*. That part of the material which gives the right ascensions and declinations, together with commentaries on each constellation, has been fully translated and tabulated by Tsuchihashi and Chevalier (abstract by Rigge).[52] Equatorial planispheres are included in this work. Less attention, however, has been paid to the ecliptic coordinate data, though these are of interest to us in connection with the Korean screen. As for von Hallerstein, he succeeded Kögler in the Directorship of the Board of Astronomy and Calendar, holding it for nearly thirty years. Gogeisl followed Pereira and von Hallerstein as Assistant Director; this post he held for twenty-six years. One of the quadrants in the Peking Observatory was probably

[50] It is ironical that only in China was it possible to realise Tycho Brahe's deathbed wish that Kepler would support and perpetuate his system. We owe this comment to Dr Nathan Sivin; see his 'Copernicus in China' (*Colloquia Copernicana*, vol. II, Toruń, 1973).

[51] We are grateful to Dr Gari Ledyard (private comm.) for the following additional information: von Hallerstein was well known to Koreans from the *Yŏn'gi* 燕記 ('Yenching [i.e. Peking] memoir') of Hong Taeyong 洪大容, whose eighteenth-century private observatory we mentioned in Ch. 3, p. 113 above.

Hong and a Korean colleague had a long conversation, stretching over several sessions, with von Hallerstein and his aide Gogeisl in 1766. Astronomical instruments were among the topics discussed. For the text of Hong's memoir, see *Tamhŏn Yŏn'gi* 湛軒燕記, esp. pp. 240ba–245aa, repr. in *Yŏnhaengnok sŏnjip* 燕行錄選集 (Collection of selected records of travels to Yenching) (Seoul, 1960), I: 231–430. Dr Ledyard has kindly made available to us a draft translation of this document.

Dr Donald Baker of the University of Washington has also been at work on these matters, particularly the efforts of Korean Neo-Confucian scientists to come to grips with European cosmology as taught by the Jesuits. We are indebted to Dr Baker for making available to us a draft version of his article, 'Jesuit Science through Korean Eyes', now being prepared for publication.

[52] P. Tsuchihashi and S. Chevalier, 'Catalogue d'étoiles fixes observées à Pékin sous l'empereur Kien-Long', *Annalles de l'Observatoire Astronomique de Zô-sè*, Shanghai, 1914 (1911), 7. 4; English summary by W. F. Rigge in *Popular Astronomy*, 1915, *23*: 29–32.

designed by him. Da Rocha was more of a geographer and explorer than an astronomer, but he worked well in the star-catalogue team none the less.[53]

The only other Jesuit mentioned in the screen inscriptions is the lay brother Fernando Bonaventura Moggi (Li Po-ming 利博明 or Li Pai-ming 利白明, 1694–1761).[54] He was a Florentine who had studied painting, sculpture, and architecture. He built churches in China and was always willing to carry out fine inscriptions in stone or wood.

The Korean folding-screen (*wei-p'ing* 圍屛) planispheres were by no means the first of their kind to incorporate results of the astronomical work of the Jesuit Mission in China. In the year 1633 the Jesuit astronomer Johann Adam Schall von Bell (T'ang Jo-wang) made for the emperor a splendid map of the heavens in eight sections suitable for mounting as a wall diagram or on a screen. He was assisted by another Jesuit, the Italian Giacomo Rho (Lo Ya-ku 羅雅谷, 1593–1638), and by a number of Chinese scholars and astronomers, including the famous Hsü Kuang-ch'i 徐光啓, Wu Ming-chu 鄔明著, and eight junior observers. The prefatory inscription was written by Hsü Kuang-ch'i only a few months before his death late in 1633, and the large sheets of Chinese paper must have been printed off from wood blocks early in the following year. Although the title is *Ch'ih-tao nan-pei liang tsung hsing-t'u* 赤道南北兩總星圖 (Equatorial Star-Charts of the Southern and Northern Hemispheres), the ecliptic poles and their circles of celestial longitude are also shown. The constellations are Chinese, not European, except for the southern part of the southern hemisphere, for which the Jesuits brought new information. In the corners between the large diagrams there are twelve further smaller discoidal diagrams, some containing graphs of the movements and retrogradations of the planets, others giving auxiliary star-maps. In addition there are four small square pictures showing astronomical apparatus, two armillary spheres, one quadrant, and one sextant. Working from two copies preserved in the Vatican Library, Pasquale d'Elia has given a description of the whole production, together with a translation of the preface written by Hsü Kuang-ch'i and the explanations by J. A. Schall von Bell.[55] Just as in our Korean screen, the northern hemisphere is on the right and the southern hemisphere on the left.

Although extremely rare, these planispheres of J. A. Schall von Bell and Hsü

[53] He became the eighth European Director of the Board.

[54] Pfister, *Notices*, no. 313.

[55] Pasquale M. d'Elia, 'The Double Stellar Hemisphere of Johann Schall von Bell S.J. (Peking, 1634)', *Monumenta Serica*, 1959, *18*: 328–59.

Kuang-ch'i are not unique. One of us (JN) had the opportunity of examining other copies and similar charts in the possession of Mr Philip Robinson of London in 1956. One double planisphere in a set of six sheets was very similar to the eight-sheet production described by d'Elia but had only six smaller discoidal diagrams inserted in the spandrels. (The subsequent identification of two further sheets led to the realisation that the item is a near duplicate (except that it is uncoloured) of the one described by d'Elia. Of these two sheets one, signed by J. A. Schall von Bell, has three planetary disc diagrams and a picture of a Ptolemaic ecliptic armillary sphere wrongly captioned 'equatorial'. There is also a picture of a sextant. The other, signed by Hsü Kuang-ch'i, also shows three planetary disc diagrams and an equatorial armillary sphere – a direct copy of the lesser armillary of Tycho Brahe – wrongly captioned 'ecliptic'. It also shows an altazimuth quadrant. The two sheets also contain the dedicatory text by Hsü Kuang-ch'i and the explanatory text by Schall von Bell.)[56] Another is a smaller chart in scroll form giving planispheres with ecliptic coordinates only, and therefore approximating more to the Korean screen. There is no date or main title, but the explanatory text has the title *Huang-tao nan-pei liang tsung hsing-t'u shuo* 黃道南北兩總星圖說 (Descriptions of the Star-Charts of the Southern and Northern Hemispheres on Ecliptic Coordinates). This text has the signature of J. A. Schall von Bell and a statement that it was written by Chu Mou-yuan 祝懋元 and collated by Ch'en Ying-teng 陳應登. It also reproduces in print the seal of the Jesuit order. There was also a smaller scroll of two hemispherical equatorial planispheres entitled *Ch'ih-tao nan-pei liang tsung hsing-t'u*. This has, within a small table of star-magnitude symbols, the signature of Nan Huai-jen (Ferdinand Verbiest); it may be dated to about 1672.[57] In addition there were twenty or thirty loose proof-sheets of pictures of astronomical instruments, the first one having pasted on it a small piece of paper xylographically printed with the words '*observationes Astronomicae*'.[58]

[56] These two sheets were discussed as a separate item in the 1966 article by Needham and Lu (of which this chapter is a revision). Unbeknown to them, some time after JN's 1956 visit Robinson had identified the two sheets as belonging with the set of six. See Philip Robinson, 'Collector's Piece VI: "Phillipps 1986", The Chinese Puzzle', *Book Collector*, Summer 1976, *25*.2: 171–94, pp. 188–90.

[57] Carlos Sommervogel, *Bibliothèque de la Compagnie de Jésus: Bibliographie* (Paris, 1890–1932), VIII: col. 578, no. 18; Pfister, *Notices*, no. 17.

[58] The print with the piece of paper pasted on it has been identified (JHC, unpublished) as the first of the nineteen figures ((v) and (vi) below) of Verbiest's *Compendium latinum* . . . (see (ii) and (iii) below). It has been verified that eleven of the others seen with it in 1956 belonged to the same series, and four to the series mentioned in (vii) below.

Note 59 of the 1966 article by Needham and Lu was written after a preliminary examination of a folio-sized xylographic volume of works by Verbiest in the library of the School of Oriental and African Studies, University of London (Acc. no. 35409), which had led to an incorrect ascription of the prints to Verbiest's *Astronomia Europaea* ... (cf. (iv) below). After further investigation, the contents of that volume have been identified as follows:

(i) Latin title-page *Liber Organicus* ... (cf. Sommervogel, col. 576, no. 6; Pfister, *Notices*, no. 37), xylographed *c.* 1678 to be prefixed to copies of Verbiest's *I-hsiang t'u* 1674 intended for Western recipients. Dated M.DC.LXVIII by evident mistake for M.DC.LXXVIII.

(ii) Text of six pages, but no title-page, of the first part of Verbiest's *Compendium latinum proponens XII posteriores figuras Libri Observationum Nec non VII priores figuras Libri Organici* (Sommervogel, col. 576, no. 7; cf. Pfister, *Notices*, no. 38). This text reappears, edited by Couplet with minor amendments, as the first part (pp. 40–5) of Ch. XII in his edition of *Astronomia Europaea* ... The wording of the heading to this chapter repeats that of the *Compendium latinum* ... title-page, but without the word 'latinum', and with 'VIII' substituted for 'VII'.

(iii) Text, of eleven pages, of the second part of Verbiest's *Compendium latinum* ... This text reappears, edited by Couplet with minor amendments, as the second part (pp. 45–57) of Ch. XII in his edition of *Astronomia Europaea* ... This part begins with a sub-heading the wording of which repeats that of the *Liber Organicus* title-page ((i) above), but with its first line altered to read 'Compendium libri organici ...'

(iv) Latin title-page *Astronomia Europaea* ... (cf. Sommervogel, col. 576, No. 8), xylographed *c.* 1678 from the MS. title-page to Verbiest's draft of the autobiographical account published by Couplet under this title at Dillingen in 1687 (Sommervogel, col. 580, no. 24; Pfister, *Notices*, No. 36). Dated M.DCLXVIII by evident mistake for M.DCLXXVIII. The Latin text to which this title-page belongs is not present, and may never have been xylographed, but appears edited as Chs. I–XI of the 1687 edition.

(v) The twelve prints referred to in the first part of the title, and first part of the text, of Verbiest's *Compendium latinum* ... ((ii) above). The first print, which shows the use of a gnomon and its shadow-scale for determining solar elevation, is a duplicate of the one referred to in the first paragraph of this note, p. 173, and has a duplicate piece of paper pasted on it. (The print in question has been reproduced by H. Bosmans, 'Ferdinand Verbiest, Directeur de l'Observatoire de Péking', Société Scientifique de Bruxelles, *Revue des Questions Scientifiques*, *71*: 195–273, 375–464, Fig. 2, and by A. Damry, 'Le p. Verbiest et l'astronomie Sino-Européenne', *Ciel et Terre: Bulletin de la Société Belge d'Astronomie* (Brussels), 1913, *34.7*: 215–39, Fig. 18.)

(vi) The seven prints referred to in the second part of the title, and described (with some mention of further plates which are lacking) in the second part of the text (item (iii) above), of Verbiest's *Compendium latinum* ... These seven prints are duplicates of the first 7 of the 105 prints in (vii) below, and they differ (together with the next 4) from the remainder in bearing Chinese titles and ordinal prefixes to their figure-numbers. The first print (unnumbered) is the view of the re-equipped Peking Observatory that was re-engraved by Melchior Haffner for the frontispiece of the 1687 edition of *Astronomia Europaea* ... (reproduced in Needham, SCC III: Fig. 190); the copy in the set of seven has pasted on it a piece of paper xylographically printed with the words '*Compendium Astronomiae Organicae*'.

(vii) Preface in Chinese, and the 105 prints, but no title-page, of Verbiest's (*Hsin-chih*) (*ling-t'ai*) *i-hsiang t'u* (Pfister, *Notices*, No. 9; cf. Sommervogel, col 575, No. 4). The first print, and the individual drawings (numbered 1–117 in Chinese) which appear on the remaining plates, are as reprinted (from new wood-blocks) in the *T'u-shu chi-ch'eng*, *li-fa tien*, ch. 93–5 (Needham, SCC III: 452, n. e). The *c.* 50,000-character explanatory text, chuan 1–4 of (*Hsin-chih*) (*ling-t'ai*) *i-hsiang chih* (Pfister, *Notices*, No. 8; cf. Sommervogel, col. 575, No. 3), is not present in the SOAS volume, but is reprinted in *Li-fa tien*, ch. 89–92, and contains many references to individual figures. According to Jeon, STK, p. 31, Verbiest's *I-hsiang chih* was reprinted in Korea in 1714. The early part of the Chinese text has a Latin counterpart in the second part of *Compendium latinum* ... ((iii) above). A Latin version of the whole Chinese text may never have been written or xylographed, but Chs. XIII–XXVIII of the 1687 edition of *Astronomia Europaea* are an extensive autobiographical account of the circumstances in which the various parts of the Chinese text were written and illustrated.

Three years later, in 1959, Mr Robinson consulted one of us (JN) again about an engraved star-map which had come to light; this (reproduced in Fig. 5.6) was none other than the original of the Jesuit planispheres and the upper inscription (dated 1723) painted on the Korean screen. There are only slight differences in wording in the two versions of the inscription. Mr Robinson informed us that the engraving here published (with his permission) appeared among a number of original documents and holograph letters of Antoine Gaubil, S.J. (Sung Chün-jung 宋君榮, 1689–1759), the famous historian of Chinese astronomy, especially those which he addressed to E. Souciet, S.J., his editor at the Jesuit College in Paris *c.* 1725. One of Gaubil's letters encloses a letter, written in Latin by Kögler on 13 March 1726, which refers to enclosures including this very star-map. The MS. notes in Latin seen on the engraving, then quite new, are therefore most probably in Kögler's own hand; they include the usual Western cursive symbols for the planets. The drawings on the screen correspond well enough with those on the engraving, save that the screen shows little or no attempt to reproduce the surface features of the moon found on the engraving. It is indeed satisfying to know that this example of the original Kögler–Moggi engraving still exists.[59]

In the early 1960s, an eight-panel screen bearing a star-map on ecliptic coordinates, executed in Korea but based on the Kögler 1723 planisphere, was discovered in the Pŏpchu Temple 法住寺 in Korea. This screen, dated 1743, is similar in many respects to the screen described in this chapter, though its inscriptions are less comprehensive. In the course of research into the screen's origins, an example of the Kögler engraved planisphere was discovered in the files of the National Museum at Seoul. This example of the original star-map had presumably been in the possession of the Yi court, forgotten for many decades.[60]

THE BACKGROUND OF SINO-KOREAN RELATIONS

Korean relations with China and the Jesuit astronomers there went back quite far in the history of modern science. Early in 1631 a Korean embassy headed by the

[59] There is another example in the Bibliothèque Nationale, Paris. The engraving is listed in Pfister's bibliography of Kögler's works, *Notices*, p. 647.

[60] Yi Yongbŏm 李龍範 [Lee Yongbum], 'Pŏpchusa sojang ŭi sinpŏp ch'ŏnmun tosŏl e taehayŏ' 法住寺所藏의新法天文圖說에 대 하여 (On the astronomical map of 1743 preserved in the Pŏpchu Temple), *Yŏksa hakpo* 歷史學報 (Korean Historical Review), 1966, *31*: 1–66; *32*: 59–119; English summary pp. 196–8. See also Jeon, STK, pp. 29–30, and the Japanese edition of STK (*Kankoku kagaku gijutsu shi*, Tokyo, 1978), pp. 38–41 and Fig. 1–8.

scholar Chŏng Tuwŏn 鄭斗源 arrived in Peking. On the way they had disembarked at Tengchow 登州 in Shantung where at that time the governor, Sun Yuan-hua 孫元化, was a Christian (he was called Ignatius). Through this contact they had met a Portuguese priest, João Rodrigues, S.J. (Lu Jo-han 陸若漢, 1561–1634),[61] who was very much interested in astronomy, though he is not usually regarded as having been one of the Jesuit astronomers, and he was not indeed a member of the China Mission, because he had come from the Japanese province. This priest was commonly known by the sobriquet Tcuzzu or T'ung-shih 通事 because of his skill as an interpreter. In the previous year he had come north from Macao with an artillery detachment commanded by Captain Concalo Teixeira-Correa (Kung-sha Ti-hsi-lao (Hsiao-chung) 公沙的西勞 (効忠)) to give help to the Ming Dynasty against the threatening Manchu invasion. Teixeira-Correa lost his life in the following year (1632) when the Ming troops under Sun Yuan-hua revolted; Rodrigues, however, escaped and returned with honours to Macao. A letter from Rodrigues to Chŏng Tuwŏn emphasising the advances of modern astronomy has been preserved and may be read in a translation by d'Elia.[62]

The result of this contact was that when Chŏng Tuwŏn returned to Korea later in 1631 he took with him many books and instruments; these probably included all the tractates and treatises which the Jesuits had published in Chinese on astronomy and calendrical science down to that time. The three monographs which dealt with the telescope and its results would certainly not have been omitted. For Chŏng Tuwŏn also took back with him a telescope (*ch'ien-li ching* 千里鏡) with which objects could be seen a hundred *li* 里 away, including 'even the smallest things in the enemy's camp'.[63] This was a personal gift from Rodrigues himself. It is interesting that the terrestrial use of the telescope in warfare loomed larger in this description than its employment in astronomy. It is also interesting that just at this very time spotting telescopes were being used in battle in Central China, far away to the south.[64] Chŏng Tuwŏn also took away, besides a volume on the manners and customs of Western countries, a treatise on cannon and their use, and he took a small field-gun itself which could be fired by a flint without

[61] Pfister, *Notices*, no. 71 (p. 214, corrected as on pp. 23*ff).

[62] D'Elia, *Galileo*, pp. 42ff.

[63] Jeon, STK, p. 77.

[64] See the study by Joseph Needham and Lu Gwei-djen, 'The "Optick Artists" of Chiangsu', *Proceedings of the Royal Microscopical Society*, 1967, *2* (part 1): 113–38.

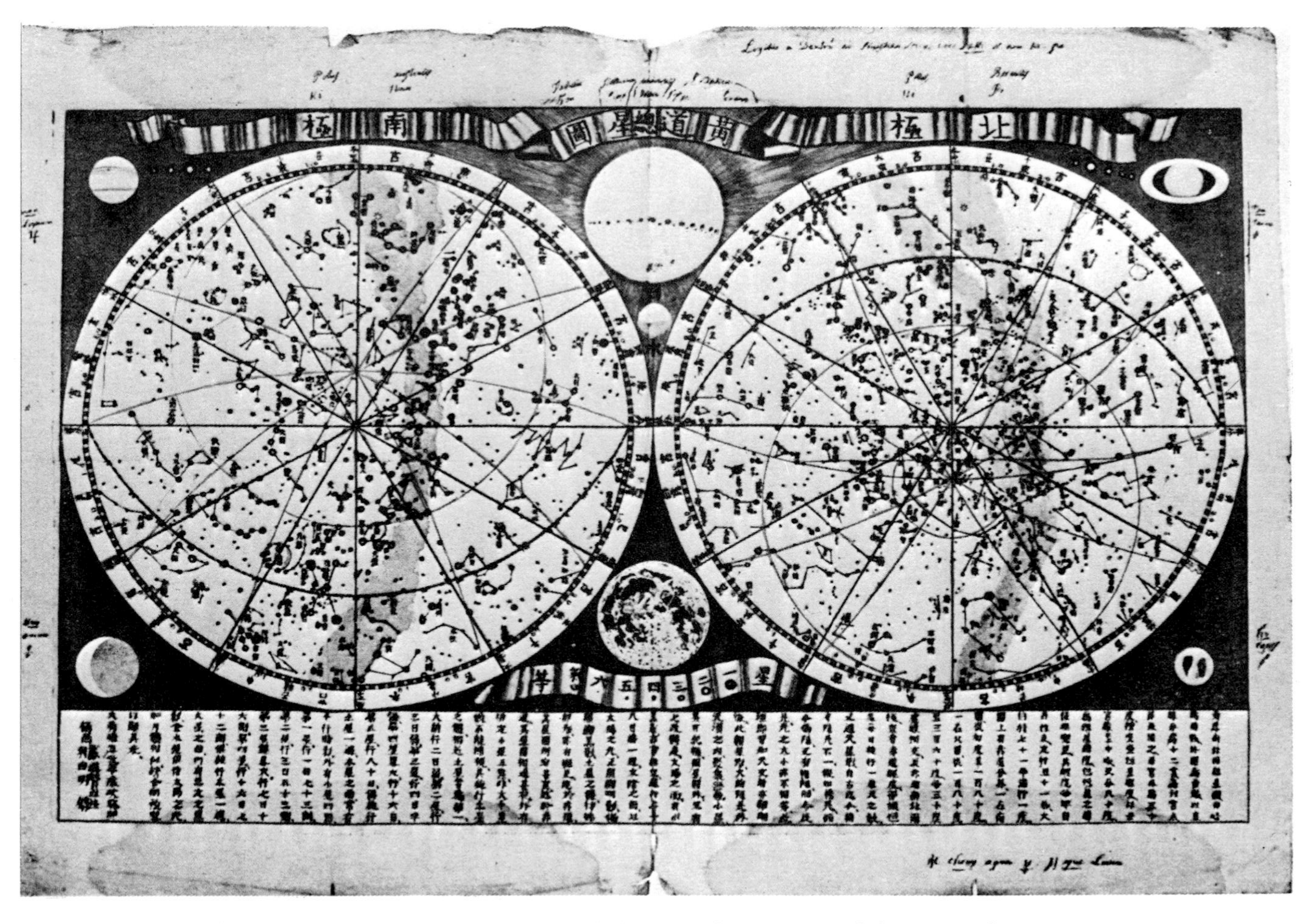

Fig. 5.6. The engraved star-map of the ecliptic planispheres, with diagrams of sun, moon, and planets, issued by Ignatius Kögler and Fernando Bonaventura Moggi in 1723; the original of the main part of the Jesuit contribution to the Korean astronomical screen.

need for a fuse. There were also some European clocks and scientific instruments in his baggage, and last but not least there was a large astronomical chart. For all this information we are indebted to a long passage in the Korean historical work *Kukcho pogam* 國朝寶鑑 (Precious Mirror of the Dynasty).[65]

Similar contacts continued as time went on. In 1644, just after the Manchus had taken over the throne of China, a close friendship developed between the young Korean crown prince in exile and the astronomer J. A. Schall von Bell. When the prince finally returned home he took with him presents from Schall von Bell, including works on science and religion, a map of the world, and a celestial globe.[66] In 1645 the Director of the Korean Bureau of Astronomy, Kim Yuk 金堉, was asked by the king to reform the calendar, so he sent a younger colleague, Han Hŭng'il 韓興一, to Peking, where he obtained books on modern astronomy.[67] In 1648 another Korean, Song Inyŏng 宋仁龍, studied modern methods under Schall von Bell, and returned with an astronomical chart ten feet long. In 1653 another Korean, Kim Sangbŏm 金尚范, studied also in Peking.[68] And there were probably many other such contacts which further research would reveal.[69]

Fortunately, we have another record of this time just when we need it to explain the Korean screen that is the subject of this chapter. According to the *Chŭngbo munhŏn pigo*,[70] the Korean embassy to China in 1741 included two interpreter-secretaries, An Kukpin 安國麟 and Pyŏn Chunghwa 卞重和. During their time in Peking they formed a close friendship with the Jesuits Ignatius

[65] Begun by Kwon Nam 權擥 in the middle of the fifteenth century and continued as the official history of the Yi Dynasty. On Chŏng's visit see Rufus, 'Astronomy in Korea', p. 26, and d'Elia, *Galileo*, pp. 42ff.

[66] Rufus, 'Astronomy in Korea', p. 37.

[67] Jeon, STK, p. 83 (Jeon reports that Kim Yuk himself went to Peking in 1644).

[68] Rufus, 'Astronomy in Korea', p. 38; Jeon, STK, pp. 83–4.

[69] On the other hand, we recognise the importance of not reading too much 'scientific exchange' into isolated incidents. In the published article on which this chapter is based, the following note appeared at this point: 'Scientific interests were reciprocated; for example the Ming admiral Mao Wen-lung 毛文龍 obtained Korean calendrical and astronomical writings by special request in 1625.' Dr Gari Ledyard (private comm.) has kindly shown us that this misrepresents the nature of the situation. Admiral Mao, he points out, was operating in Korean waters; he needed a new calendar, requested one from the Korean authorities, and eventually obtained one (*Injo sillok*, 8: 7a). The main issue involved in this incident was a political rather than a scientific one. Korea, which in acknowledging Ming suzerainty also accepted the official Ming calendar, could not legally (under the terms of its 'tributary' status) print calendars on its own. Yet as a practical matter it had to do so, and did. What was then to be done when a *Ming* official asked the *Korean* court for a calendar? The matter was debated at court in 1625; in the end it was decided to send Admiral Mao the calendar, while apologising for the apparent violation of Ming law and excusing the illegal printing on the grounds of practical exigency.

[70] Cf. Rufus, 'Astronomy in Korea', p. 43.

Kögler and André Pereira. From these astronomers they obtained ephemerides of the sun, moon, and planets, tables for computation (probably logarithms), lists of solar and lunar eclipses, and treatises on mathematics. The copy of the Kögler planisphere now in the Seoul National Museum may well have been amongst these documents. The Pŏpchu Temple astronomical screen (1743) no doubt was based on this star-map. It will be noted also that the sojourn in China of An and Pyŏn came only fourteen years before the termination of the Ch'ien-lung star-catalogue with its planispheres. We may well feel justified in thinking that the friendship of these four men was kept up after the Koreans returned to their own country; that would have provided a perfect route of transmission for the up-to-date information in the inscriptions of the Korean screen described here.

It only remains to allude to a curious historical coincidence. The screen that we have described rests now in the Whipple Museum of the History of Science at Cambridge University. It was the wish of its previous owner, Mr Chang Chung-ti, that it should come to Cambridge, for his grandfather, Mr Minn Kiusik, had been one of the first Koreans to study there; he was an undergraduate at Trinity Hall. But the English connection goes back much further. For the Jesuit astronomer in Peking, André Pereira, was not really a Portuguese, as his name would imply; his real name was Andrew Jackson, and he was the only Englishman who ever served in the noble ranks of the Jesuit Mission in China. He was born at Oporto, probably in a family concerned with the wine trade, and upon being naturalised and joining the Latin Church he took a Portuguese name. Thus we may feel that the permanent home of the Korean screen in Cambridge commemorates not only a recent personal association but also the eighteenth-century friendship of An Kukpin and Andrew Jackson in Peking.

EPILOGUE

Both the adherence to traditional East Asian astronomy and the introduction of new ideas and techniques from the West continued in Korea for another 150 years after the middle of the eighteenth century, during the long twilight of the Yi Dynasty. These latter-day events have been well surveyed by Jeon.[1] The remarkable instruments that we have discussed in these pages are products of a truly Korean tradition of astronomy and technology, formed in large part from, but not always limited by, influences from China and, later, from the Peking Jesuits and from Japan. They reflect, but also in some ways transcend, the variety of their origins. The eighteenth-century astronomical screen described in the preceding chapter is an appropriate note on which to end our account, for in many ways it typifies Korean astronomy for the remainder of the Yi period.

At the official level of the Royal Observatory, Jesuit knowledge and techniques had wrought irreversible changes, and to that extent Korean astronomy was slowly making its way into the international mainstream of science. Yet Jesuit astronomy was itself conservative (and cosmologically misleading),[2] and it was grafted onto an even more conservative Korean base. Many more decades would have to elapse before Korea was exposed fully to the diversity and rapid evolution of modern science.

Twice, at the beginning and during the middle years of the Yi Dynasty, Korean astronomers (like, indeed, everyone else at the Korean court) had to respond to the political and intellectual shock of dynastic overthrow in China; the supersession of traditional Korean astronomy coincided with the downfall of China's last dynasty. After the fall of the Ch'ing in 1911 there were neither a new dynastic calendar, renovation of instruments, nor reinvigorated cosmological speculation. Traditional Chinese science had become quite out of date by the late nineteenth century, and the successors of the Ch'ing state abandoned its vestiges in their

[1] Jeon, STK, *passim*.

[2] Sivin, 'Copernicus in China'.

urgent need to catch up with the West. It would be years before China could again claim a position of prominence in science – this time, the emerging ecumenical science of the twentieth century.

Korea, in any case, was in no position to take its lead from the Republican heirs of the Ch'ing in 1911. The end of Korea's tributary status was presaged by the Kwanghwa Treaty of 1876 between Korea and Japan, and the age-old relationship between Korea and China was ended two decades later by the Sino-Japanese War. The Yi Dynasty preceded the Ch'ing into extinction by one year, in 1910, and Korea was jolted into the modern world by thirty-five years of harsh occupation by the rapidly modernising Japanese. The Royal Observatory was no more, and Korean astronomy became only the province of popular astrologers and almanac-makers.

The survival, through all these upheavals, of the Song Iyŏng / Yi Minch'ŏl armillary clock, the eighteenth-century astronomical screens, and a few other instruments and other relics of the Yi Royal Observatory provides us with a reminder of that proud tradition of astronomy, and with precious historical evidence of the integration of traditional East Asian science into the one ever evolving science of the modern world.

APPENDIX: THE KINGS OF THE YI DYNASTY TO 1776

T'aejo	太祖	(1392–8)
Chŏngjong	定宗	(1398–1400)
T'aejong	太宗	(1400 – 18)
Sejong	世宗	(1418–50)
Munjong	文宗	(1450–2)
Tanjong	端宗	(1452–5)
Sejo	世祖	(1456–68)
Yejong	睿宗	(1468–9)
Sŏngjong	成宗	(1469–94)
Yonsan-gun	燕山君	(1494–1506)
Chungjong	中宗	(1506–44)
Injong	仁宗	(1544–5)
Myŏngjong	明宗	(1545–67))
Sŏnjo	宣祖	(1567–1608)
Kwanghae-gun	光海君	(1608–23)
Injo	仁祖	(1623–49)
Hyojong	孝宗	(1649–59)
Hyŏnjong	顯宗	(1659–74)
Sukchong	肅宗	(1674–1720)
Kyŏngjong	景宗	(1720–4)
Yŏngjo	英祖	(1724–76)

SELECT BIBLIOGRAPHY

von Bertele, Hans. *Globes and Spheres*. Lausanne: Scriptar S.A., 1961.

Boodberg, Peter A. 'Chinese Zoographic Names as Chronograms'. *Harvard Journal of Asiatic Studies*, 1940, 5: 128–36.

Bosmans, Henri. 'Ferdinand Verbiest, Directeur de l'Observatoire de Péking'. Société Scientifique de Bruxelles, *Revue des Questions Scientifiques*, *71*: 195–273, 375–464.

Bruin, Frans, and Margaret Bruin. 'The Limits of Accuracy of Aperture-Gnomons', in Y. Maeyama and W. G. Saltzer, eds., *Prismata: Naturwissenschaftsgeschichtliche Studien. Festschrift für Willy Hartner*. Wiesbaden: Franz Steiner Verlag, 1977, pp. 21–42.

Chang Chia-t'ai 张家泰. 'Teng-feng kuan-hsing-t'ai ho Yüan-ch'u t'ien-wen kuan-ts'e ti ch'eng-chiu' 登封观星台和元初天文观测的成就 (On the observation tower at Teng-feng and the achievements of astronomical observation in the early Yüan period). *K'ao-ku* 考古, 1976.2: 95–102; repr. in *Chung-kuo t'ien-wen-hsüeh shih wen-chi* (*q.v.*), pp. 229–41.

Ch'ao t'ien lu – Ming-tai Chung-Han kuan-hsi shih-liao hsuan-chi 朝天錄～明代中韓關係史料選輯 (Daily court records – historical materials on the relations between China and Korea during the Ming). Taipei, 1978.

Ch'en, Kenneth. 'Matteo Ricci's Contribution to, and Influence on, Geographical Knowledge in China'. *Journal of the American Oriental Society*, 1939, *59*: 325–59, errata 509.

Choi Hyon Pae (for the King Seijong Memorial Society). *King Seijong the Great*. Seoul, 1970.

Chu K'o-chen 竺可桢. 'Lun ch'i-yü chin-t'u yü han-tsai' 论祈雨禁屠与旱灾 (A discussion of praying for rain (by) prohibiting the slaughter of animals, in connection with drought-disasters). Originally published in *Tung-fang tsa-chih* 東方雜誌 (The Eastern Miscellany), 1926, *23*. 13: 5–18; repr. in *Chu K'o-chen wen-chi* 竺可桢文集 (Collected works of Chu K'o-chen). Peking: Science Press, 1979, pp. 90–9.

Chŭngbo munhŏn pigo 增補文獻備考 (Comprehensive study of [Korean] civilisation, revised and expanded), 1790, 1908. Officially compiled. Modern repr. edn, Seoul: *Kosŏ Kan-haenghoe* 古書刊行會, 1959. Enlarged from the *Tongguk munhŏn pigo*

東國文獻備考 (Study of the civilisation of the Eastern Kingdom), 1770, officially compiled by Hong Ponghan 洪鳳漢 *et al.*

Chung-kuo t'ien-wen-hsüeh shih wen-chi 中国天文学史文集 (Collected essays in the history of Chinese astronomy). Peking, 1978.

CLUTTON, C., AND G. DANIELS. *Clocks and Watches in the Collection of the Worshipful Company of Clockmakers*. London: Sotheby Parke Bernet Publications, 1975; repr. 1980.

COMBRIDGE, JOHN H. 'The Astronomical Clocktowers of Chang Ssu-hsun and his Successors, A.D. 976 to 1126'. *Antiquarian Horology*, June 1975, *9*.3: 288–301.

'The Celestial Balance: A Practical Reconstruction'. *Horological Journal*, Feb. 1962, *104*.2: 82–6.

'Chinese Sexagenary Calendar-Cycles'. *Antiquarian Horology*, Sept. 1966, *5*.4: 134.

'Chinese Steelyard Clepsydras'. *Ibid.* Spring 1981, *12*.5: 530–5.

'Clockmaking in China: Early History', in Alan Smith (ed.), *The Country Life International Dictionary of Clocks*. Feltham, Middx, 1979.

'Clocktower Millenary Reflections'. *Antiquarian Horology*, Winter 1979, *11*.6: 604–8.

'Hour Systems in China and Japan'. *Bulletin of the National Association of Watch and Clock Collectors, Inc.*, Aug. 1976, *18*.4: 336–8.

CULLEN, CHRISTOPHER. 'Joseph Needham on Chinese Astronomy'. *Past and Present*, 1980 *87*: 39–53.

'Some Further Points on the *Shih*'. *Early China*, 1980–1, *6*: 31–46.

CULLEN, CHRISTOPHER, AND ANN S. L. FARRER, 'On the Term *Hsüan Chi* and the Flanged Trilobate Jade Discs'. *Bulletin of the School of Oriental and African Studies*, 1983, *46*: 52–76.

DAMRY, A. 'Le p. Verbiest et l'astronomie sino-européenne'. *Ciel et Terre: Bulletin de la Société Belge d'Astronomie* (Brussels), 1913, *34*.7: 215–39.

D'ELIA, PASQUALE M. 'The Double Stellar Hemisphere of Johann Schall von Bell S.J. (Peking, 1634)'. *Monumenta Serica*, 1959, *18*: 328–59.

Galileo in Cina: relazioni attraverso il Collegio Romano tra Galileo e i gesuiti scienzati missionari in Cina (1610–1640). Rome, 1947. English tr. with emendations and additions by R. Suter and M. Sciascia, *Galileo in China*. Cambridge, Mass.: Harvard University Press, 1960.

Erh-shih-ssu shih 二十四史 (The twenty-four [officially compiled dynastic] histories). *Po-na-pen* 百納本 edn. Shanghai and Taipei: Commercial Press, various reprints.

FRANKE, HERBERT. 'Beiträge z. Kulturgeschichte Chinas unter der Mongolenherrschaft'. *Abhandlungen für die Kunde des Morgenlandes*, 1956, *32*: 1–160.

HARPER, DONALD. 'The Han Cosmic Board (*shih* 式)'. *Early China*, 1978–9, *4*: 1–10.

'The Han Cosmic Board: A Response to Christopher Cullen'. *Early China*, 1980–1, *6*: 47–56.

HATADA TAKASHI. *A History of Korea*, tr. W. W. Smith, Jr, and B. H. Hazard. Santa Barbara: University of California (Santa Barbara) Press, 1969.

HAZARD, BENJAMIN H., *et al. Korean Studies Guide*. Berkeley: University of California Press, 1954.

HENTHORN, WILLIAM E. *A History of Korea*. New York: Free Press, and London: Collier-Macmillan, 1971; repr. 1974.

HIGGINS, KATHLEEN. 'The Classification of Sundials'. *Annals of Science*, 1953, *9*: 342–58.

HILL, DONALD R. *Arabic Water-Clocks*. Aleppo, Syria: University of Aleppo Institute for the History of Arabic Science, 1981.

HILL, DONALD R., ed. and tr. *On the Construction of Water-Clocks: Kitāb Arshimīdas fī ʿamal al-binkamāt*. (Occasional Paper No. 4.) London: Turner and Devereux, 1976.

HILL, H.O., AND E. W. PAGET-TOMLINSON. *Instruments of Navigation*. London: National Maritime Museum, 1958.

HO PENG YOKE [HO PING-YÜ]. *The Astronomical Chapters of the Chin Shu*. Paris and The Hague: Mouton, 1966.

HONG ISŎP 洪以燮. *Chosŏn kwahaksa* 朝鮮科學史 (History of Korean science). Seoul, 1946.

HONG TAEYONG 洪大容. *Tamhŏnso* 湛軒書 (Works of Tamhŏn [= Hong Taeyong]). Repr. edn, Seoul, 1939.

Hsin i-hsiang fa-yao 新儀象法要 (New design for an astronomical clock), by Su Sung 蘇頌, 1092; ed. Shih Yuan-chih 施元之, 1172. Repr. (4to) 1844, (8vo) 1889, 1922 (in *Shou-shan-ko ts'ung-shu* 守山閣叢書), 1935–7 (in *Ts'ung-shu chi-ch'eng* 叢書集成), and 1969 (Taipei, *Jen-jen wen-k'u* 人人文庫, no. 1248).

HULBERT, HOMER B. *History of Korea*. 2 vols., Seoul, 1905. Ed. Clarence N. Weems as *Hulbert's History of Korea*, 2 vols. London, 1962.

HUMMEL, ARTHUR W., ed. *Eminent Chinese of the Ch'ing Period (1644–1912)*, 2 vols. Washington: Library of Congress, 1943–4.

I SHIH-T'UNG 伊世同. 'Liang-t'ien-ch'ih k'ao' 量天尺考 (On the 'foot' used in celestial measurements). *Wen-wu* 文物, 1978.2: 10–17.

'Cheng-fang an k'ao' 正方案考 (On the True-Direction Table), *Wen-wu*, 1982.1: 76–7,82.

JEON SANG-WOON (Chŏn Sangun 全相運). '15 segi chŏnban Yijo kwahak kisulsa sŏsŏl' 15 世紀前半李朝科學技術史序說 (An introduction to the history of science and technology in early 15th-century Korea), in *Ilsan Kim Tujong paksa hŭisu kinyŏm nonmunjip* 一山金斗鐘博士稀壽記念論文集 (Commemorative papers for Dr Kim Tujong's seventieth birthday). Seoul, 1966.

'Han'guk ch'ŏnmun kisanghaksa' 韓國天文氣象學史 (A history of astronomy and meteorology in Korea), in *Kwahak kisulsa* 科學技術史 (History of science and technology), in the *Han'guk munhwasa taegang* 韓國文化史大綱 (History of Korean culture) series. Seoul, 1968.

Han'guk kwahak kisulsa 韓國科學技術史 (A history of science and technology in Korea). Seoul, 1966. 2nd Korean edn, with revisions and additional illustrations, 1976. *See also* Jeon Sang-won, *Science and Technology in Korea.*

Kankoku kagaku gijutsu shi 韓国科学技术史. Tokyo, 1978. (Japanese edn of *Science and Technology in Korea*).

'Meteorology in the Yi Dynasty, Korea'. *Theses Collection of Sungshin Women's Teachers College*, 1968, *1*: 61–75.

'Richō jidai ni okeru kōuryō sokuteihō ni tsuite' 李朝時代における降雨量測定法について (On the scientific measure of precipitation in the Yi Dynasty). *Kagakushi kenkyū* 科學史研究, 1963, *66*: 49–56.

Science and Technology in Korea: Traditional Instruments and Techniques. Cambridge, Mass. and London: M.I.T. Press, 1974. Rev. and tr. from Jeon, *Han'guk kwahak kisulsa*, 1966, *q.v.*

'Sŏn'gi okhyŏng e taehayŏ' 璿璣玉衡어對하여 (On armillary spheres with clockwork in the Yi Dynasty). *Ko munhwa* 古文化, 1963, *2*: 2–10. Same article summarised in Japanese, with English résumé, 'Senki gyokkō (tenmon tokei) ni tsuite' 璇璣玉衡(天文時計)について, in *Kagakushi kenkyū*, 1962, *63*: 137–41.

'Sŏun Kwan kwa kanŭidae' 書雲觀과簡儀臺. (The Bureau of Astronomy and the observatory in the Yi Dynasty). *Hyangt'o sŏul* 鄕土서울, 1964, *20*: 37–51.

'Understanding of Science in History of Korea, with Emphasis on the Scientists in the Early 15th Century'. *Japanese Studies in the History of Science*, 1967.6: 124–37.

'Yissi Chosŏn ŭi sigye chejak sogo' 李氏朝鮮의時計製作小考 (A study of timekeeping instruments in the Yi Dynasty). *Hyangt'o sŏul*, 1963, *17*: 49–114.

KIM, TU-JONG. *A Bibliographical Guide to Traditional Korean Sources.* Seoul: Asiatic Research Centre, Korea University, 1976.

KIM YANGSŎN 金良善. 'Myŏngmal Ch'ŏngch'o Yasohoe sŏn'gyosa tŭri chejak han segye chido wa kŭ Han'guk munhwasasang e mich'in yŏnghyang' 明末淸初耶蘇會宣教師들이製作한世界地圖와그韓國文化史上에이친影響 (Jesuit world maps of the late Ming and early Ch'ing, and their influence in Korea). *Sungdae* 崇大, 1961, *6*: 16–58.

KIM YUK 金堉. *Chamgok p'iltam* 潛谷筆談 (Miscellaneous writings of Chamgok [= Kim Yuk]), in *Chamgok chŏnjip* 潛谷全集 (Complete works of Chamgok). Seoul: Taedong munhwa yŏn'guso repr. edn, 1965.

KIM, YUNG-SIK. 'The World-View of Chu Hsi (1130 to 1200): Knowledge about the Natural World in the *Chu Tzu Ch'üan Shu*'. Unpub. doctoral thesis, Princeton University, 1979.

KING, HENRY C. *Geared to the Stars.* Toronto, 1978.

KUNO, YOSHI S. *Japanese Expansion on the Asiatic Continent*. 2 vols. Berkeley, 1937–40.

KWON HYOGMYON. *Basic Chinese-Korean Character Dictionary*. Wiesbaden: Otto Harrassowitz, 1978.

[LEE SANGBAEK]. 'The Origin of Korean Alphabet "Hangul"' (English summary by Dugald Malcolm printed posthumously as Part I of) Ministry of Culture and Information, Republic of Korea, *A History of Korean Alphabet and Moveable Types*. Seoul, n.d.

LI NIEN 李儼. *Chung suan shih lun-ts'ung* 中算史論叢 (Collected essays in the history of Chinese mathematics). 2nd series, 5 vols. Peking: Science Press, 1954.

LI TI 李迪. *Kuo Shou-ching* 郭守敬. Shanghai, 1966.

Liu ching t'u 六經圖 (Illustrations of the six classics), by Yang Chia 楊甲, *c*. 1160. *Liu ching t'u ting pen* 六經圖定本, 1740 edn.

LORCH, RICHARD P. 'Al-Khāzinī's "Sphere That Rotates by Itself"', *Journal for the History of Arabic Science*, 1980, *4*: 287–329.

'Al-Khāzinī's Balance-clock and the Chinese Steelyard Clepsydra', *Archives Internationales d'Histoire des Sciences*, June 1981, *31*: 183–9.

LYONS, H. G. 'An Early Korean Rain-Gauge'. *Quarterly Journal of the Royal Meteorological Society*, 1924, *50*: 26.

MCCUNE, GEORGE M. AND EDWIN O. REISCHAUER. *The Romanisation of the Korean Language Based upon its Phonetic Structure*. Repr. Seoul, n.d. from *Transactions of the Korea Branch of the Royal Asiatic Society*, 1939, *29*: 1–55.

MCCUNE, SHANNON B. 'Old Korean World Maps'. *Korean Review*, 1949, *2.1*: 14–17.

MAJOR, JOHN S. 'Myth, Cosmology, and the Origins of Chinese Science'. *Journal of Chinese Philosophy*, 1978, *5*: 1–20.

'New Light on the Dark Warrior', in N. J. Girardot and J. S. Major, eds., *Myth and Symbol in Chinese Tradition*. Boulder, Colo.: *Journal of Chinese Religions* symposium volume, 1985.

MARUYAMA KIYOYASU 丸山清康. 'Hōken shakai to gijutsu – wadokei ni shūyaku sareta hōken gijutsu' 封建社会と技術〜和時計に集約された封建技術 (Feudal society and technology – feudal technology revealed by Japanese clocks). *Kagakushi kenkyū*, Sept. 1954, 31: 16–22.

MASPERO, HENRI. 'Les instruments astronomiques des chinois aux temps des Han'. *Mélanges Chinois et Bouddhiques*, 1938–9, *6*: 183–370.

NAKAMURA, H. 'Old Chinese World Maps Preserved by the Koreans'. *Imago Mundi: A Revue of Early Cartography*, 1947, *4*: 3–22.

'Old Chinese World Map Preserved by the Koreans'. *Chōsen gakuhō* 朝鮮學報, 1966, *39–40*: 1–73. (Correction to preceding article.)

NAKAYAMA, SHIGERU. *A History of Japanese Astronomy: Chinese Background and Western Impact*. Cambridge, Mass.: Harvard University Press, 1969.

NEEDHAM, JOSEPH. 'Astronomy in Ancient and Medieval China'. *Philosophical Transactions of the Royal Society, London*, 1974, ser. A, *276*: 67–82.

'The Peking Observatory in A.D. 1280 and the Development of the Equatorial Mounting'. *Vistas in Astronomy*, 1955, *1*: 67–83.

Science and Civilisation in China. 7 vols. projected. Cambridge: Cambridge University Press, 1954–. Vol. III, 1959, and vol. IV.2, 1965.

NEEDHAM, JOSEPH, AND LU GWEI-DJEN. 'A Korean Astronomical Screen of the Mid-Eighteenth Century from the Royal Palace of the Yi Dynasty (Chosŏn Kingdom, 1392 to 1910)'. *Physis*, 1966, *8*.2: 137–62.

'The "Optick Artists" of Chiangsu'. *Proceedings of the Royal Microscopical Society*, 1967, *2* (part 1): 113–38. Abstract, *ibid.* 1966, *1* (part 2): 59–60.

NEEDHAM, JOSEPH, WANG LING, AND D. J. DE SOLLA PRICE. *Heavenly Clockwork: The Great Astronomical Clocks of Medieval China*. Cambridge: Cambridge University Press, 1960; 2nd edn, with new Foreword and Supplement, 1986.

NELTHROPP, H. L. *Catalogue of the Nelthropp Collection*. 2nd edn, London, 1900.

NODA CHŪRYŌ 能田忠亮. *Tōyō tenmongaku-shi ronsō* 東洋天文學史論叢 (Collected papers on the history of astronomy in East Asia). Tokyo, 1944.

P'AN NAI 潘鼐. 'Nan-ching ti liang-t'ai ku-tai ts'e-t'ien i-ch'i – Ming-chih hun-i ho chien-i' 南京的两台古代测天仪器～明制浑仪和简仪 (Two ancient observational astronomical instruments at Nanking – the armillary sphere and Simplified Instrument made during the Ming period). *Wen-wu* 1975.7: 84–9.

'Suchou Nan-Sung t'ien-wen t'u-p'ai ti k'ao-shih yü p'i-p'an' 苏州南宋天文图碑的考释与批判 (Study and critique of the Southern Sung planisphere stele at Suchou). *K'ao-ku hsüeh-pao* 考古学报, 1976.1: 47–61. English summary, *ibid.*: 62.

PARK SONG-RAE. 'Portents and Politics in Early Yi Korea, 1392–1519'. Unpub. doctoral thesis, University of Hawaii, 1977.

PFISTER, L. *Notices biographiques et bibliographiques sur les Jésuites de l'ancienne mission de Chine* (1552 to 1773). 2 vols. (Variétés Sinologiques, nos. 59–60). Shanghai: Mission Press, 1932–4.

PRICE, DEREK J. 'Clockwork before the Clock'. *Horological Journal*, 1955, *97*: 810–14; 1956, *98*: 31–5.

'A Collection of Armillary Spheres and Other Antique Scientific Instruments'. *Annals of Science*, 1954, *10*: 172–87.

'The Prehistory of the Clock'. *Discovery*, April 1956, *17*.4: 153–7.

Richō jitsuroku 李朝實錄 (Veritable records of the Yi Dynasty, 1418–1864). Repr. edn with index and supplementary materials. Tokyo: Gakushūin Institute of Oriental Culture, 1953.

RIGGE, W. F. 'A Chinese Star-Map Two Centuries Old'. *Popular Astronomy*, 1915, *23*: 29–32.

ROBERTSON, J. DRUMMOND. *The Evolution of Clockwork*. London, 1931; repr. Wakefield: S. R. Publishers, 1972.

ROBINSON, PHILIP. 'Collector's Piece VI: "Phillipps 1986", The Chinese Puzzle'. *Book Collector*, Summer 1976, *25*.2: 171–94.

(ROBINSON, T. O.). 'A Korean 17th Century Armillary Clock' (notes on a lecture by J. H. Combridge on 27 November 1964). *Antiquarian Horology*, March 1965, *4*: 300–1.

R(OBINSON), T. O. 'A Transitional Japanese Clock'. *Antiquarian Horology*, 1966, *5*: 97.

RUFUS, W. CARL 'Astronomy in Korea'. *Transactions of the Korea Branch of the Royal Asiatic Society*, 1936, *26*: 1–52.

'The Celestial Planisphere of King Yi Tai-jo'. *Ibid*. 1913, *4*.3: 23–72.

'Korea's Cherished Astronomical Chart'. *Popular Astronomy*, 1915, *23*.4: 193–8.

RUFUS, W. CARL, AND CELIA CHAO. 'A Korean Star Map'. *Isis*, 1944, *35*: 316–26.

RUFUS, W. CARL, AND WON-CHUL LEE. 'Marking Time in Korea'. *Popular Astronomy*, 1936, *44*: 252–7.

RUFUS, W. CARL, AND HSING-CHIH TIEN. *The Soochow Astronomical Chart*. Ann Arbor: University of Michigan Press, 1945.

DE SAUSSURE, LÉOPOLD. *Les origines de l'astronomie chinoise*. Paris, 1930. Repr., with review by A. Pogo, Taipei, 1967.

SCHLEGEL, GUSTAVE. *Uranographie chinoise*. 2 vols. Leiden: E. J. Brill, 1875; repr. Taipei, 1967.

SHINJŌ SHINZŌ 新城新藏. *Tōyō tenmongaku-shi kenkyū* 東洋天文學史研究 (Researches on the history of astronomy in East Asia). Tokyo, 1929.

Shu-chuan ta-ch'uan 書傳大全 (Complete commentaries on the Book of Documents). Repr. Korea, *c*. 1620.

SIVIN, NATHAN. 'Copernicus in China', in Union Internationale d'Histoire et de philosophie des Sciences, Comité Nicolas Copernic, ed., *Colloquia Copernicana*, II: *Études sur l'audience de la théorie heliocentrique*. Conférences du Symposium de l'UIHPS, Toruń, 1973. Warsaw *et al*., 1973, pp. 63–122.

Cosmos and Computation in Early Chinese Mathematical Astronomy. Leiden: E. J. Brill, 1969.

SMITH, ALAN, ed. *The Country Life International Dictionary of Clocks*. Feltham, Middx, 1979.

SOHN POW-KEY, KIM CHOL-CHOON, AND HONG YI-SUP. *The History of Korea*. Seoul: Korean National Commission for UNESCO, 1970.

SOMMERVOGEL, CARLOS. *Bibliothèque de la Compagnie de Jésus: Bibliographie*, vol. 8. Brussels and Paris, 1893.

SONG SANG-YONG. 'A Brief History of the Study of the Ch'ŏmsŏng-dae in Kyŏngju'. *Korea Journal*, Aug. 1983, *23*.8: 16–21.

STEVENSON, EDWARD L. *Terrestrial and Celestial Globes: Their History and Construction.* 2 vols. New Haven: Yale University Press for Hispanic Society of America, 1921.

SYMONDS, R(OBERT) W. *A History of English Clocks.* Harmondsworth and New York: Penguin, 1947. 2nd edn, *A Book of English Clocks*, 1950.

TAKABAYASHI HYOGO 高林兵衛. *Tokei hattatsu shi* 時計發達史 (A history of the development of timekeepers). Tokyo, 1924.

Tokei no hanashi 時計の話 (The story of timekeepers). Tokyo, 1925.

TAMURA SENNOSUKE 田村專之助. 'Chōsen Richō gakusha no chikyū kaitensetsu ni tsuite' 朝鮮李朝學者の地珠回轉說について (On the rotating-earth theory of Yi Dynasty scholars). *Kagakushi kenkyū*, July 1954, 30: 23–4.

TSUCHIHASHI, P., AND S. CHEVALIER. 'Catalogue d'étoiles fixes observées à Pékin sous l'empereur Kien-Long', *Annales de L'Observatoire Astronomique de Zô-sè*, Shanghai, 1914 (1911), 7.4. English summary by W. F. Rigge in *Popular Astronomy*, 1915, *23*: 29–32.

TURNER, ANTHONY J. *The Time Museum Catalogue, Volume 1, Part 3: Water-clocks, Sand-glasses, Fire-clocks.* Rockford, Illinois: The Time Museum, 1984.

WADA, YUJI. 'A Korean Rain-Gauge of the 15th Century'. *Quarterly Journal of the Royal Meteorological Society*, 1911, *37*: 83–6.

WANG CHIEN-MIN 王建民, *et al.* 'Tseng Hou-i mu ch'u-t'u ti erh-shih-pa hsiu ch'ing-lung pai-hu t'u-hsiang' 曾侯乙墓出土的二十八宿青龙白虎图象 (On a picture of the twenty-eight Lunar Lodges, the Blue-Green Dragon, and the White Tiger, excavated from the tomb of the Marquis Yi of Tseng). *Wen-wu*, 1979.7: 40–5.

WARD, F. A. B. *Time Measurement, Part 1: Historical Review.* 1st edn, London, H.M.S.O., 1936: many later edns.

WHITE, W. C. AND P. M. MILLMAN. 'An Ancient Chinese Sun-Dial'. *Journal of the Royal Astronomical Society of Canada*, Nov. 1938, *32*.3: 417–30.

WYLIE, ALEXANDER. *Notes on Chinese Literature.* Shanghai and London, 1867. 2nd edn, Shanghai, 1901; repr. 1902. 3rd edn, Shanghai, 1922; repr. Peking: Vetch, 1939.

Chinese Researches. Shanghai, 1897; repr. Taipei, 1966.

XI ZEZONG [HSI TSE-TSUNG] 席泽宗. 'Chinese Studies in the History of Astronomy, 1949–1979'. *Isis*, Sept. 1981, *72*: 456–70.

XI ZEZONG [HSI TSE-TSUNG] AND BO SHUREN [PO SHU-JEN] 薄树人. 'Chung Ch'ao Jih san-kuo ku-tai ti hsin-hsing chi-lu chi ch'i tsai she-tien t'ien-wen-hsüeh chung ti i-i' 中朝日三国古代的新星纪录及其在射电天文学中的意义. *T'ien-wen hsüeh-pao* 天文學報 (*Acta Astronomia Sinica*), 1965, *13*.1: 1–22. Tr. as S. R. Bo and Z. Z. Xi, 'Ancient Novae and Supernovae Recorded in the Annals of China, Korea and Japan and their Significance in Radio Astronomy'. NASA TT-F-388 (Technical Translations Series), 1966.

YABUUCHI KIYOSHI 藪内清. 'Chūgoku no takei' 中国の時計 (Timekeeping instruments in ancient China). *Kagakushi kenkyū*, July 1951, 19: 19–22.

'Richō gakusha no chikyū kaitensetsu' 李朝學者の地転説 (On the rotating-earth theory of scholars in the Yi Dynasty). *Chōsen gakuhō* 朝鮮學報, 1968, *49*: 427–34.

YAMADA KEIJI 山田慶児. *Jujireki no michi* 授時暦の道 (The principles of the Shou-shih calendrical system). Tokyo: Mizusu Shobo, 1980.

'Kōdai no mizudokei' 古代の水時計 (Ancient water-clocks). *Shizen* 自然, March 1983: 58–66.

YAMAGUCHI MASAYUKI 山口正之. 'Shinchō ni okeru zai-Shi Ōjin to Chōsen shishin' 清朝に於ける在支歐人と朝鮮使臣 (Jesuits and Korean envoys in Ch'ing China). *Shigaku zasshi* 史學雜誌, 1933, *44*.7: 1–30 (*44*: 795–824).

'Shōken Seshi to Tō Jakubō' 昭顯世子と湯若望 (On Prince Sohyŏn and Adam Schall von Bell). *Seikyū gakusō* 青丘學叢, 1931, *5*: 112–17.

YAMAGUCHI RYŪJI 山口隆二. *Nihon no tokei* 日本の時計 (The clocks of Japan). Tokyo, 1942. 2nd edn, rev. with English introduction as *The Clocks of Japan*, Tokyo, 1950.

YANG YU 楊瑀. *Shan-chu hsin-hua* 山居新話 (Conversations in the mountain retreat concerning recent events), *c*. 1360.

YI KYŎNGCH'ANG 李慶昌. *Chuch'ŏn tosol* 周天圖說 (Illustrated explanation of the Revolving Heaven theory), *c*. 1601.

YI YONGBŎM 李龍範 [Lee Yongbum]. 'Pŏpchusa sojang ŭi sinpŏp ch'ŏnmun tosŏl e taehayŏ' 法住寺所藏의新法天文圖說에 대하여 (On the astronomical map of 1743 preserved in the Pŏpchu Temple). *Yŏksa hakpo* 歷史學報 (Korean Historical Review), 1966, *31*: 1–66; *32*: 59–119; English summary, pp. 196–8.

YOSHIDA MITSUKUNI 吉田光邦. 'Kongi to konshō' 渾儀と渾象 (Celestial globes and armillary spheres [in China]). *Silver Jubilee Volume of the Zinbun Kagaku Kenkyuso, Kyoto University*. Kyoto, 1954.

YU YŎNGBAK 유영박. 'Sejong ŭi sahoe chŏngch'aek' 世宗의社會政策 (Social policies of King Sejong). *Chindan hakpo* 震檀學報, 1966, *29/30*: 129–44. Part of *Tugye paksa kohŭi kinyŏm nonmunjip* 斗溪博士古稀記念論文集 (Commemorative papers for Dr Yi Pyŏngdo's seventieth birthday).

YULE, Sir HENRY. *The Book of Ser Marco Polo*. 3rd edn, rev. Henri Cordier. 2 vols. London, 1903; repr. 1921.

INDEX

Page numbers in *italics* refer to illustrations or their captions.

INDEX

INDEX